ISAAC ASIMO foremost write An Associate the Boston Uni he has written well over a hundred books, as well as hundreds of articles in publications ranging from *Esquire* to Atomic Energy Commission pamphlets. Famed for his science fiction writing (his three-volume Hugo Award-winning THE FOUNDATION TRILOGY is available in individual Avon editions and as a one-volume Equinox edition), Dr. Asimov is equally acclaimed for such standards of science reportage as THE UNIVERSE, LIFE AND ENERGY, THE SOLAR SYSTEM AND BACK, ASIMOV'S BIOGRAPHICAL ENCYCLOPEDIA OF SCIENCE AND TECHNOLOGY, and ADDING A DIMENSION (all available in Avon editions). His non-science writings include the two-volume ASIMOV'S GUIDE TO SHAKESPEARE, ASIMOV'S ANNOTATED DON JUAN, and the two-volume ASIMOV'S GUIDE TO THE BIBLE (available in a two-volume Avon edition). Born in Russia, Asimov came to this country with his parents at the age of three, and grew up in Brooklyn. In 1948 he received his Ph.D. in Chemistry at Columbia and then joined the faculty at Boston University, where he works today.

ADDING
A DIMENSION

ISAAC ASIMOV

 DISCUS BOOKS/PUBLISHED BY AVON

To the Gentle Readers,
who have always been as
gentle as any man could wish.

AVON BOOKS
A division of
The Hearst Corporation
959 Eighth Avenue
New York, New York 10019

Copyright © 1964 by Isaac Asimov.
Published by arrangement with Doubleday and Company, Inc.

ISBN: 0-380-00278-7

First Discus Printing, March, 1975.

DISCUS TRADEMARK REG. U.S. PAT. OFF. AND
FOREIGN COUNTRIES, REGISTERED TRADEMARK—
MARCA REGISTRADA, HECHO EN CHICAGO, U.S.A.

Printed in Canada

CONTENTS

Introduction

A NUMBER of years ago, when I was a freshly appointed instructor, I met, for the first time, a certain eminent historian of science. At the time I could only regard him with tolerant condescension.

I was sorry for a man who, it seemed to me, was forced to hover about the edges of science. He was compelled to shiver endlessly in the outskirts, getting only feeble warmth from the distant sun of science-in-progress; while I, just beginning my research, was bathed in the heady liquid heat at the very center of the glow.

In a lifetime of being wrong at many a point, I was never more wrong. It was I, not he, who was wandering in the periphery. It was he, not I, who lived in the blaze.

I had fallen victim to the fallacy of the "growing edge"; the belief that only the very frontier of scientific advance counted; that everything that had been left behind by that advance was faded and dead.

But is that true? Because a tree in spring buds and comes greenly into leaf, are those leaves therefore the tree? If the newborn twigs and their leaves were all that existed, they would form a vague halo of green suspended in mid air, but surely that is not the tree. The leaves, by themselves, are no more than trivial fluttering decoration. It is the trunk and limbs that give the tree its grandeur and the leaves themselves their meaning.

There is not a discovery in science, however revolutionary, however sparkling with insight, that does not arise out of what went before. "If I have seen further than other men," said Isaac Newton, "it is because I have stood on the shoulders of giants."

And to learn that which goes before does not detract from the beauty of a scientific discovery but, rather, adds to it; just as the gradual unfolding of a flower, as seen by time-lapse photography, is more wonderful than the mature flower itself, caught in stasis.

In fact, an overly exclusive concern with the growing edge can kill the best of science, for it is not on the growing edge itself that growth can best be seen. If the growing edge only is studied, science begins to seem a revelation without a history of development. It is Athena, emerging adult and armed from the forehead of Zeus, shouting her fearful war cry with her first breath.

How dare one aspire to add to such a science? How can one ward off bitter disillusion when part of the structure turns out to be wrong. The perfection of the growing edge is meretricious while it exists, hideous when it cracks.

But add a dimension!

Take the halo of leaves and draw it together with branches that run into limbs that join to form a trunk that firmly enters the ground. It is the tree of science that you will then see, an object that is a living, growing, and permanent thing; not a flutter of leaves at the growing edge, insubstantial, untouchable, and dying with the frosts of fall.

Science gains reality when it is viewed not as an abstraction, but as the concrete sum of work of scientists, past and present, living and dead. Not a statement in science, not an observation, not a thought exists in itself. Each was ground out of the harsh effort of some man, and unless you know the man and the world in which he worked; the assumptions he accepted as truths; the concepts he considered untenable; you cannot fully understand the statement or observation or thought.

Consider some of what the history of science teaches.

First, since science originated as the product of men and not as a revelation, it may develop further as the continuing product of men. If a scientific law is not an eternal truth but merely a generalization which, to some man or group of men, conveniently described a set of observations, then to some other man or group of men, another generalization might seem even more convenient. Once it is grasped that scientific truth is limited and not absolute, scientific truth becomes capable of further refinement. Until that is understood, scientific research has no meaning.

Second, it reveals some important truths about the humanity of scientists. Of all the stereotypes that have plagued men of science, surely one above all has wrought harm. Scientists can be pictured as "evil," "mad," "cold," "self-centered," "absent-minded," even "square" and yet survive easily. Unfortunately, they are usually pictured as "right" and that can distort the picture of science past redemption.

Scientists share with all human beings the great and inalienable privilege of being, on occasion, wrong; of being egregiously wrong sometimes, even monumentally wrong. What is worse still, they are sometimes perversely and persistently wrong-headed. And since that is true, science itself can be wrong in this aspect or that.

With the possible wrongness of science firmly in mind, the student of science today is protected against disaster. When an individual theory collapses, it need not carry with it one's faith and hope and innocent joy. Once we learn to expect theories to collapse and to be supplanted by more useful generalizations, the collapsing theory becomes not the gray remnant of a broken today, but the herald of a new and brighter tomorrow.

Third, by following the development of certain themes in science, we can experience the joy and excitement of the grand battle against the unknown. The wrong turnings, the false clues, the elusive truth nearly captured half a century before its time, the unsung prophet, the false authority, the hidden assumption and cardboard syllogism, all add to the suspense of the struggle and make what we slowly gain through the study of the history of science worth more than what we might quickly gain by a narrow glance at the growing edge alone.

To be sure, the practical thought might arise: But would it not be better if we learned the truth at once? Would we not save time and effort?

Yes, we might, but it is not as important to save time and effort as to enjoy the time and effort spent. Why else should a man rise before dawn and go out in the damp to fish, waiting happily all day for the occasional twitch of his line when, without getting out of bed, he might have telephoned the market and ordered all the fish he wanted?

It is for this reason, then, that I present this new collection of essays. It is my hope that, every once in a while, some vignette of Science Past may illuminate some corner of Science Present.

Part One

MATHEMATICS

1.   T-Formation

I HAVE been accused of having a mad passion for large numbers and this is perfectly true. I wouldn't dream of denying it. However, may I point out that I am not the only one?

For instance, in a book entitled *Mathematics and the Imagination* (published in 1940) the authors, Edward Kasner and James Newman, introduced a number called the "googol," which is good and large and which was promptly taken up by writers of books and articles on popular mathematics.

Personally, I think it is an awful name, but the young child of one of the authors invented it, and what could a proud father do? Thus, we are afflicted forever with that baby-talk number.

The googol was defined as the number 1 followed by a hundred zeros, and so here (unless I have miscounted or the Noble Printer has goofed) is the googol, written out in full:
10.000.000.000.000.000,000,000,000,000.000,000,000,000,
000,000,000.000,000,000,000,000,000,000,000,000,000,000,
000,000,000,000,000,000,000.000.

Now this is a pretty clumsy way of writing a googol, but it fits in with our system of numeration, which is based on the number 10. To write large numbers we simply multiply 10's, so that a hundred is ten times ten and is written 100; a thousand is ten times ten times ten and is written

1000 and so on. The number of zeros in the number is equal to the number of tens being multiplied, so that the googol, with a hundred zeros following the 1, is equal to a hundred tens multiplied together. This can also be written as 10^{100}. And since 100 is ten times ten or 10^2, the googol can even be written as 10^{10^2}.

Certainly, this form of exponential notation (the little figure in the upper right of such a number is an "exponent") is very convenient, and any book on popular math will define a googol as 10^{100}. However, to anyone who loves large numbers, the googol is only the beginning and even this shortened version of writing large numbers isn't simple enough.[1]

So I have made up my own system for writing large numbers and I am going to use this first chapter as a chance to explain it. (Freeze, everyone! No one's leaving till I'm through.)

The trouble, it seems to me, is that we are using the number 10 to build upon. That was good enough for cave men, I suppose, but we moderns are terribly sophisticated and we know lots better numbers than that.

For instance, the annual budget of the United States of America is in the neighborhood, now, of $100,000,000,000 (a hundred billion dollars). That means 1,000,000,000,000 (one trillion) dimes.

Why don't we, then, use the number, one trillion, as a base? To be sure, we can't visualize a trillion, but why should that stop us? We can't even visualize fifty-three. At least if someone were to show us a group of objects and tell us there are fifty-three of them altogether, we couldn't tell whether he were right or wrong without counting them. That makes a trillion no less unreal than fifty-three, for we have to count both numbers and both are equally countable. To be sure, it would take us much longer to count one trillion than to count fifty-three, but the principle is the same and I, as anyone will tell you, am a man of principle.

The important thing is to associate a number with something physical that can be grasped and this we have done.

---

[1] The proper name for the googol, before I forget, is "ten duotrigintillion," but I dare say, gloomily, that that will never replace "googol."

The number 1,000,000,000,000 is roughly equal to the number of dimes taken from your pocket and mine (mostly mine, I sometimes sullenly think) each year by kindly, jovial Uncle Sam to build missiles and otherwise run the government and the country.

Then, once we have it firmly fixed in our mind as to what a trillion is, it takes very little effort of imagination to see what a trillion trillion is; a trillion trillion trillion, and so on. In order to keep from drowning in a stutter of trillions, let's use an abbreviated system that, as far as I know, is original with me.[2]

Let's call a trillion T-1; a trillion trillion T-2; a trillion trillion trillion T-3, and form large numbers in this fashion. (And there's the "T-formation" of the title! Surely you didn't expect football?)

Shall we see how these numbers can be put to use? I have already said that T-1 is the number of dimes it takes to run the United States for one year. In that case, T-2 would represent the number of dimes it would take to run the United States for a trillion years. Since this length of time is undoubtedly longer than the United States will endure (if I may be permitted this unpatriotic sentiment) and, in all likelihood, longer than the planet earth will endure, we see that we have run out of financial applications of the Asimovian (ahem!) T-numbers long before we have even reached T-2.

Let's try something else. The mass of any object is proportional to its content of protons and neutrons which, together, may be referred to as nucleons. Now T-1 nucleons make up a quantity of mass far too small to see in even the best optical microscope and even T-2 nucleons make up only 1⅔ grams of mass, or about $\frac{1}{16}$ of an ounce.

Now we've got room, it would seem, to move way up the T-scale. How massive, for instance, are T-3 nucleons? Since T-3 is a trillion times as large as T-2, T-3 nucleons have a mass of 1.67 trillion grams, or a little under two

<hr>

[2]Actually, Archimedes set up a system of numbers based on the myriad, and spoke of a myriad myriad, a myriad myriad myriad and so on. But a myriad is only 10,000 and I'm using 1,000,000,-000,000, so I don't consider Archimedes to be affecting my originality. Besides, he only beat me out by less than twenty-two centuries.

million tons. Maybe there's not as much room as we thought.

In fact, the T-numbers build up with breath-taking speed. T-4 nucleons equals the mass of all the earth's ocean, and T-5 nucleons equals the mass of a thousand solar systems. If we insist on continuing upward, T-6 nucleons equal the mass of ten thousand galaxies the size of ours, and T-7 nucleons are far, far more massive than the entire known universe.

Nucleons are not the only subatomic particles there are, of course, but even if we throw in electrons, mesons, neutrinos, and all the other paraphernalia of subatomic structure, we cannot reach T-7. In short, there are far less than T-7 subatomic particles of all sorts in the visible universe.

Clearly, the system of T-numbers is a powerful method of expressing large numbers. How does it work for the googol? Well, consider the method of converting ordinary exponential numbers into T-numbers and vice versa. T-1 is equal to a trillion, or 10^{12}; T-2 is equal to a trillion trillion, or 10^{24}, and so on. Well, then, you need only divide an exponent by 12 to have the numerical portion of a T-number; and you need only multiply the numerical portion of a T-number by 12 to get a ten-based exponent.

If a googol is 10^{100}, then divide 100 by 12, and you see at once that it can be expressed as T-8½. Notice that T-8½ is larger than T-7 and T-7 is in turn far larger than the number of subatomic particles in the known universe. It would take a billion trillion universes like our own to contain a googol of subatomic particles.

What then is the good of a googol, if it is too large to be useful in counting even the smallest material objects spread through the largest known volume?

I could answer: For its own sheer, abstract beauty—

But then you would all throw rocks at me. Instead, then, let me say that there are more things to be counted in this universe than material objects.

For instance, consider an ordinary deck of playing cards. In order to play, you shuffle the deck, the cards fall into a certain order, and you deal a game. Into how many different orders can the deck be shuffled? (Since it is impossible to have more essentially different game-situations than there are orders-of-cards in a shuffled deck, this is a question that should interest your friendly neighborhood poker-player.)

The answer is easily found (if you know where to look, and I do) and comes out to 80,000,000,000,000,000,000,000,000,000,000,000,000,000,000,000,000,000,000,000,000,000,000,000,000,000,000.000, or 8×10^{67}. In T-numbers, this is something like T-5⅔. With an ordinary deck of cards, then, we can count arrangements and reach a value equal to that of the number of subatomic particles in a galaxy, more or less.

If, instead of 52 cards, we played with 70 cards (and this is not unreasonable; canasta, I understand, uses 108 cards), then the number of different orders after shuffling, just tops the googol mark.

So when it comes to analyzing card games (let alone chess, economics, and nuclear war), numbers like the googol and beyond are met with.

Mathematicians, in fact, are interested in many varieties of numbers (with and without practical applications) in which vastnesses far, far beyond the googol are quickly reached.

Consider Leonardo Fibonacci, for instance, the most accomplished mathematician of the Middle Ages. (He was born in Pisa, so he is often called Leonardo of Pisa.) About 1200, when Fibonacci was in his prime, Pisa was a great commercial city, engaged in commerce with the Moors in North Africa. Leonardo had a chance to visit that region and profit from a Moorish education.

The Moslem world had by that time learned of a new system of numeration from the Hindus. Fibonacci picked it up and in a book, *Liber Abaci*, published in 1202, introduced these "Arabic numbers" and passed them on to a Europe still suffering under the barbarism of the Roman numerals. (Since Arabic numerals are only about a trillion times as useful as Roman numerals, it took a mere couple of centuries to convince European merchants to make the change.)

In this same book Fibonacci introduces the following problem: "How many rabbits can be produced from a single pair in a year if every month each pair begets a new pair, which from the second month on become productive, and no deaths occur?" (It is also assumed that

15

each pair consists of a male and female and that rabbits have no objection to incest.)

In the first month we begin with a pair of immature rabbits, and in the second month we still have one pair, but now they are mature. By the third month they have produced a new pair, so there are two pairs, one mature, one immature. By the fourth month the immature pair has become mature and the first pair has produced another immature pair, so there are three pairs, two mature and one immature.

You can go on if you wish, reasoning out how many pairs of rabbits there will be each month, but I will give you the series of numbers right now and save you the trouble. It is:

$$1, 1, 2, 3, 5, 8, 13, 21, 34, 55, 89, 144$$

At the end of the year, you see, there would be 144 pairs of rabbits and that is the answer to Fibonacci's problem.

The series of numbers evolved out of the problem is the "Fibonacci series" and the individual numbers of the series are the "Fibonacci numbers." If you look at the series you will see that each number (from the third member on) is the sum of the two preceding numbers.

This means we needn't stop the series at the twelfth Fibonacci number (F_{12}). We can construct F_{13} easily enough by adding F_{11} and F_{12}. Since 89 and 144 are 233, that is F_{13}. Adding 144 and 233 gives us 377 or F_{14}. We can continue with F_{15} equal to 610, F_{16} equal to 987, and so on for as far as we care to go. Simple arithmetic, nothing more than addition, will give us all the Fibonacci numbers we want.

To be sure, the process gets tedious after a while as the Fibonacci numbers stretch into more and more digits and the chances of arithmetical error increases. One arithmetical error anywhere in the series, if uncorrected, throws off all the later members of the series.

But why should anyone want to carry the Fibonacci sequence on and on and on into large numbers? Well, the series has its applications. It is connected with cumulative growth, as the rabbit problem shows, and, as a matter of fact, the distribution of leaves spirally about a lengthening stem, the scales distributed about a pine cone, the seeds distributed in the sunflower center, all have an arrangement related to the Fibonacci series. The series is also related to the "golden section," which is important to art and aesthetics as well as to mathematics.

But beyond all that, there are always people who are fascinated by large numbers. (I can't explain the fascination but believe me it exists.) And if fascination falls short of working away night after night with pen and ink, it is possible, these days, to program a computer to do the work, and get large numbers that it would be impractical to try to work out in the old-fashioned way.

The October 1962 issue of *Recreational Mathematics Magazine*[3] lists the first 571 Fibonacci numbers as worked out on an IBM 7090 computer. The fifty-fifth Fibonacci number passes the trillion mark, so that we can say that F_{55} is greater than T-1.

From that point on, every interval of fifty-five or so Fibonacci numbers (the interval slowly lengthens) passes another T-number. Indeed, F_{481} is larger than a googol. It is equal to almost one and a half googols, in fact.

Those multiplying rabbits, in other words, will quickly surpass any conceivable device to encourage their multiplication. They will outrun any food supply that can be dreamed up, any room that can be imagined. There might be only 144 at the end of a year, but there would be nearly 50,000 at the end of two years, 15,000,000 at the end of three years, and so on. In thirty years there would be more rabbits than there are subatomic particles in the known universe, and in forty years there would be more than a googol of rabbits.

To be sure, human beings do not multiply as quickly as Fibonacci's rabbits, and old human beings do die. Nevertheless, the principle remains. What those rabbits can do in a few years, we can do in a few centuries or millenniums. Soon enough. Think of that when you tend to minimize the population explosion.

For the fun of it, I would like to write F_{571}, which is the largest number given in the article. (There will be larger numbers later, but I will not write them out!) Anyway, F_{571}

---

[3]This is a fascinating little periodical which I heartily recommend to any nut congruent to myself.

is: 9604120061892255382394288336092486502610491741
18770678168222647890290143783084788641925890841
85254331637646183008074629. This vast number is not
quite equal to T-10.[4]

For another example of large numbers, consider the
primes. These are numbers like 7, or 641, or 5237, which
can be divided evenly only by themselves and 1. They have
no other factors. You might suppose that as one goes higher
and higher in the scale of numbers, the primes gradually
peter out because there would be more and more smaller
numbers to serve as possible factors.

This, however, does not happen, and even the ancient
Greeks knew that. Euclid was able to prove quite simply
that if all the primes are listed up to a "largest prime," it
is always possible to construct a still larger number which
is either prime itself or has a prime factor that is larger
than the "largest prime." It follows then there is no such
thing as a "largest prime" and the number of primes is
infinite.

Yet even if we can't work out a largest prime, there is an
allied problem. What is the largest prime we know? It
would be pleasant to point to a large number and say:
"This is a prime. There are an infinite number of larger
primes, but we don't know which numbers they are. This
is the largest number we *know* to be a prime."

Once that is done, you see, then some venturesome
amateur mathematician may find a still larger prime.

Finding a really large prime is by no means easy. Earlier,
for instance, I said that 5237 is prime. Suppose you
doubted that, how would you check me? The only practical
way is to try all the prime numbers smaller than the square
root of 5237 and see which, if any, are factors. This is
tedious but possible for 5237. It is simply impractical for
really large numbers—except for computers.

Mathematicians have sought formulas, therefore, that
would construct primes. It might not give them every
prime in the book, so that it could not be used to test a
given number for prime-hood. However, it could construct
primes of any desired size, and after that the task of finding

[4]Since this was written, the editor of *Recreational Mathematics*
wrote to say that he had new Fibonacci numbers, up to F_{1000}. This
F_{1000}, with 209 digits, is something over T-17.

a record-high prime would become trivial and could be abandoned.

However, such a formula has never been found. About 1600, a French friar named Marin Mersenne proposed a formula of partial value which would occasionally, but not always, produce a prime. This formula is $2^p - 1$, where p is itself a prime number. (You understand, I hope, that 2^p represents a number formed by multiplying p two's together, so that 2^8 is $2 \times 2 \times 2 \times 2 \times 2 \times 2 \times 2 \times 2$, or 256.)

Mersenne maintained that the formula would produce primes when p was equal to 2, 3, 5, 7, 13, 17, 19, 31, 67, 127, or 257. This can be tested for the lower numbers easily enough. For instance, if p equals 3, then the formula becomes $2^3 - 1$, or 7, which is indeed prime. If p equals 7, then $2^7 - 1$ equals 127, which is prime. You can check the equation for any of the other values of p you care to.

The numbers obtained by substituting prime numbers for p in Mersenne's equation are called "Mersenne numbers" and if the number happens to be prime it is a "Mersenne prime." They are symbolized by the capital letter M and the value of p. Thus M_3 equals 7; M_7 equals 127, and so on.

I don't know what system Mersenne used to decide what primes would yield Mersenne primes in his equation, but whatever it was, it was wrong. The Mersenne numbers M_2, M_3, M_5, M_7, M_{13}, M_{17}, M_{19}, M_{31}, and M_{127} are indeed primes, so that Mersenne had put his finger on no less than nine Mersenne primes. However, M_{67} and M_{257}, which Mersenne said were primes, proved on painstaking examination to be no primes at all. On the other hand, M_{61}, M_{89}, and M_{107}, which Mersenne did not list as primes, are primes, and this makes a total of twelve Mersenne primes.

In recent years, thanks to computer work, eight more Mersenne primes have been located (according to the April 1962 issue of *Recreational Mathematics*). These are M_{521}, M_{607}, M_{1279}, M_{2203}, M_{2281}, M_{3217}, M_{4253}, and M_{4423}. What's more, since that issue, three even larger Mersenne primes have been discovered by Donald B. Gillies of the University of Illinois. These are M_{9689}, M_{9941}, and M_{11213}.

The smallest of these newly discovered Mersenne primes, M_{521}, is obtained by working out the formula $2^{521} - 1$.

You take 521 two's, multiply them together, and subtract one. The result is far, far higher than a googol. In fact, it is higher than T-13.

Not to stretch out the suspense, the largest known Mersenne prime, M_{11213}, and, I believe, the largest prime known at present, has 3375 digits and is therefore just about T-281¼. The googol, in comparison to that, is a trifle so small that there is no reasonable way to describe its smallness.

The Greeks played many games with numbers, and one of them was to add up the factors of particular integers. For instance, the factors of 12 (not counting the number itself) are 1, 2, 3, 4, and 6. Each of these numbers, but no others, will go evenly into 12. The sum of these factors is 16, which is greater than the number 12 itself, so that 12 is an "abundant number."

The factors of 10, on the other hand, are 1, 2, and 5, which yield a sum of 8. This is less than the number itself, so that 10 is a "deficient number." (All primes are obviously badly deficient.)

But consider 6. Its factors are 1, 2, and 3, and this adds up to 6. When the factors add up to the number itself, that number is a "perfect number."

Nothing has ever come of the perfect numbers in two thousand years, but the Greeks were fascinated by them, and those of them who were mystically inclined revered them. For instance, it could be argued (once Greek culture had penetrated Judeo-Christianity) that God had created the world in six days because six is a perfect number. (Its factors are the first three numbers, and not only is their sum six, but their product is also six, and God couldn't be expected to resist all that.)

I don't know whether the mystics also made a point of the fact that the lunar month is just a trifle over twenty-eight days long, since 28, with factors of 1, 2, 4, 7, and 14 (which add up to 28), is another perfect number. Alas, the days of the lunar month are actually 29½ and the mystics may have been puzzled over this slipshod arrangement on the part of the Creator.

But how many of these wonderful perfect numbers are there? Considering that by the time you reach 28, you have run into two of them, you might think there were many.

However, they are rare indeed; far rarer than almost any other well-known kind of number. The third perfect number is 496, and the fourth is 8128, and throughout ancient and medieval times, those were the only perfect numbers known.

The fifth perfect number was not discovered until about 1460 (the name of the discoverer is not known) and it is 33,550,336. In modern times, thanks to the help of the computer, more and more perfect numbers have been discovered and the total now is twenty. The twentieth and largest of these is a number with 2663 digits, and this is almost equal to T-222.

But in a way, I have been unfair to Kasner and Newman. I have said they invented the googol and I then went on to show that it was easy to deal with numbers far higher than the googol. However, I should also add they invented another number, far, far larger than the googol. This second number is the "googolplex," which is defined as equal to 10^{googol}. The exponent, then, is a 1 followed by a hundred zeros, and I could write that, but I won't. Instead, I'll say that a googolplex can be written as:

$$10^{10^{100}} \text{ or even } 10^{10^{10^2}}$$

The googol itself can be written out easily. I did it at the beginning of the article and it only took up a few lines. Even the largest number previously mentioned in this article can be written out with ease. The largest Mersenne prime, if written out in full, would take up less than two pages of this book.

The googolplex, however, cannot be written out—literally *cannot*. It is a 1 followed by a googol zeros, and this book will not hold as many as a googol zeros no matter how small, within reason, those zeros are printed. In fact, you could not write the number on the entire surface of the earth, if you made zero no larger than an atom. In fact, if you represented each zero by a nucleon, there wouldn't be enough nucleons in the entire known universe or in a trillion like it to supply you with sufficient zeros.

You can see then that the googolplex is incomparably larger than anything I have yet dealt with. And yet I can represent it in T-numbers without much trouble.

Consider! The T-numbers go up through the digits, T-1, T-2, T-3, and so on, and eventually reach T-1,000,000,000,-000. (This is a number equivalent to saying "a trillion trillion trillion trillion . . ." and continuing until you have repeated the word *trillion* a trillion times. It will take you umpty-ump lifetimes to do it, but the principle remains.) Since we have decided to let a trillion be written as T-1, the number T-1,000,000,000,000 can be written T-(T-1).

Remember that we must multiply the numerical part of the T-number by 12 to get a ten-based exponent. Therefore T-(T-1) is equal to $10^{12,000,000,000,000}$, which is more than $10^{10^{13}}$.

In the same way, we can calculate that T-(T-2) is more than $10^{10^{25}}$, and if we continue we finally find that T-(T-8) is nearly a googolplex. As for T-(T-9), that is far larger than a googolplex; in fact, it is far larger than a googol googolplexes.

One more item and I am through.

In a book called *The Lore of Large Numbers,* by Philip J. Davis, a number called "Skewes' number" is given. This number was obtained by S. Skewes, a South African mathematician who stumbled upon it while working out a complex theorem on prime numbers. The number is described as "reputed to be the largest number that has occurred in a mathematical proof." It is given as:

$$10^{10^{10^{34}}}$$

Since the googolplex is only $10^{10^{10^2}}$, Skewes' number is incomparably the greater of the two.

And how can Skewes' number be put into T-formation? Well, at this point, even I rebel. I'm not going to do it.

I will leave it to you, O Gentle Reader, and I will tell you this much as a hint. It seems to me to be obviously greater than T-[T-(T-1)].

From there on in, the track is yours and the road to madness is unobstructed. Full speed ahead, all of you.

As for me, I shall hang back and stay sane; or, at least, as sane as I ever am, which isn't much.

2. One, Ten, Buckle My Shoe

I HAVE ALWAYS BEEN taken aback a little at my inability to solve mathematical conundrums since (in my secret heart of hearts) I feel this to be out of character for me. To be sure, numerous dear friends have offered the explanation that, deep within me, there rests an artfully concealed vein of stupidity, but this theory has somehow never commended itself to me.

Unfortunately, I have no alternate explanation to suggest. You can well imagine, then, that when I come across a puzzle to which I *can* find the answer, my heart fairly sings. This happened to me once when I was quite young and I have never forgotten it. Let me explain it to you in some detail because it will get me somewhere I want to go.

The problem, in essence, is this. You are offered any number of unit weights: one-gram, two-gram, three-gram, four-gram, and so on. Out of these you may choose a sufficient number so that by adding them together in the proper manner, you may be able to weigh out any integral number of grams from one to a thousand. Well, then, how can you choose the weights in such a way as to end with the fewest possible number that will turn the trick?

I reasoned this way—

I must start with a 1-gram weight, because only by using it can I weigh out one gram. Now if I take a second 1-gram weight, I can weigh out two grams by using both 1-gram weights. However, I can economize by taking a 2-gram weight instead of a second 1-gram weight, for then not only can I weigh out two grams with it, but I can also weigh out three grams, by using the 2-gram plus the 1-gram.

What's next? A 3-gram weight perhaps? That would be wasteful, because three grams can already be weighed

23

out by the 2-gram plus the 1-gram. So I went up a step and chose a 4-gram weight. That gave me not only the possibility of weighing four grams, but also five grams (4-gram plus 1-gram), six grams (4-gram plus 2-gram), and seven grams (4-gram plus 2-gram plus 1-gram).

By then I was beginning to see a pattern. If seven grams was the most I could now reach, I would take an 8-gram weight as my next choice and that would carry me through each integral weight to fifteen grams (8-gram plus 4-gram plus 2-gram plus 1-gram). The next weight would be a 16-gram one, and it was clear to me that in order to weigh out any number of grams one had to take a series of weights (beginning with the 1-gram) each one of which was double the next smaller.

That meant that I could weigh out any number of grams from one to a thousand by means of ten and only ten weights: a 1-gram, 2-gram, 4-gram, 8-gram, 16-gram, 32-gram, 64-gram, 128-gram, 256-gram, and 512-gram. In fact, these weights would carry me up to 1023 grams.

Now we can forget weights and work with numbers only. Using the numbers 1, 2, 4, 8, 16, 32, 64, 128, 256, and 512, and those only, you can express any other number up to and including 1023 by adding two or more of them. For instance, the number 100 can be expressed as 64 plus 32 plus 4. The number 729 can be expressed as 512 plus 128 plus 64 plus 16 plus 8 plus 1. And, of course, 1023 can be expressed as the sum of all ten numbers.

If you add to this list of numbers 1024, then you can continue forming numbers up to 2047; and if you next add 2048, you can continue forming numbers up to 4095; and if you next—

Well, if you start with 1 and continue doubling indefinitely, you will have a series of numbers which, by appropriate addition, can be used to express any finite number at all.

So far, so good; but our interesting series of numbers— 1, 2, 4, 8, 16, 32, 64, . . .—seems a little miscellaneous. Surely there must be a neater way of expressing it. And there is.

Let's forget 1 for a minute and tackle 2. If we do that, we can begin with the momentous statement that 2 is 2. (Any argument?) Going to the next number, we can say that 4 is 2 times 2. Then 8 is 2 times 2 times 2; 16 is 2 times 2 times 2 times 2; 32 is . . . But you get the idea.

So we can set up the series (continuing to ignore 1) as 2, 2 times 2, 2 times 2 times 2, 2 times 2 times 2 times 2, and so on. There is a kind of pleasing uniformity and regularity about this but all those 2 times 2 times 2's create spots before the eyes. Therefore, instead of writing out all the 2's, it would be convenient to note how many 2's are being multiplied together by using the exponential method described in the previous chapter.

Thus, if 4 is equal to 2 times 2, we will call it 2^2 (two to the second power, or two squared). Again if 8 is 2 times 2 times 2, we can take note of the three 2's multiplied together by writing 8 as 2^3 (two to the third power, or two cubed). Following that line of attack we would have 16 as 2^4 (two to the fourth power), 32 as 2^5 (two to the fifth power), and so on. As for 2 itself, only one 2 is involved and we can call it 2^1 (two to the first power).

One more thing. We can decide to let 2^0 (two to the zero power) be equal to 1. (In fact, it is convenient to let any number to the zero power be equal to 1. Thus, 3^0 equals 1, and so does 17^0 and $1,965,211^0$. For the moment, however, we are interested only in 2^0 and we are letting that equal 1.)

Well, then, instead of having the series 1, 2, 4, 8, 16, 32, 64, . . . , we can have 2^0, 2^1, 2^2, 2^3, 2^4, 2^5, 2^6. . . . It's the same series as far as the value of the individual members are concerned, but the second way of writing it is prettier somehow and, as we shall see, more useful.

We can express any number in terms of these powers of 2. I said earlier that 100 could be expressed as 64 plus 32 plus 4. This means it can be expressed as 2^6 plus 2^5 plus 2^2. In the same way, if 729 is equal to 512 plus 128 plus 64 plus 16 plus 8 plus 1, then it can also be expressed as 2^9 plus 2^7 plus 2^6 plus 2^4 plus 2^3 plus 2^0. And of course, 1023 is 2^9 plus 2^8 plus 2^7 plus 2^6 plus 2^5 plus 2^4 plus 2^3 plus 2^2 plus 2^1 plus 2^0.

But let's be systematic about this. We are using ten different powers of 2 to express any number below 1024, so let's mention all of them as a matter of course. If we don't want to use a certain power in the addition that is required to express a particular number, then we need merely multiply it by 0. If we want to use it, we multiply

25

it by 1. Those are the only alternatives; we either use a certain power, or we don't use it; we either multiply it by 1 or by 0.

Using a dot to signify multiplication, we can say that 1023 is: $1 \cdot 2^9$ plus $1 \cdot 2^8$ plus $1 \cdot 2^7$ plus $1 \cdot 2^6$ plus $1 \cdot 2^5$ plus $1 \cdot 2^4$ plus $1 \cdot 2^3$ plus $1 \cdot 2^2$ plus $1 \cdot 2^1$ plus $1 \cdot 2^0$. All the powers are used. In expressing 729, however, we would have: $1 \cdot 2^9$ plus $0 \cdot 2^8$ plus $1 \cdot 2^7$ plus $1 \cdot 2^6$ plus $0 \cdot 2^5$ plus $1 \cdot 2^4$ plus $1 \cdot 2^3$ plus $0 \cdot 2^2$ plus $0 \cdot 2^1$ plus $1 \cdot 2^0$. And again, in expressing 100, we can write: $0 \cdot 2^9$ plus $0 \cdot 2^8$ plus $0 \cdot 2^7$ plus $1 \cdot 2^6$ plus $1 \cdot 2^5$ plus $0 \cdot 2^4$ plus $0 \cdot 2^3$ plus $1 \cdot 2^2$ plus $0 \cdot 2^1$ plus $0 \cdot 2^0$.

But why bother, you might ask, to include those powers you don't use? You write them out and then wipe them out by multiplying them by zero. The point is, however, that if you systematically write them all out, without exception, you can take it for granted that they are there and omit them altogether, keeping only the 1's and the 0's.

Thus, we can write 1023 as 1111111111; we can write 729 as 1011011001; and we can write 100 as 0001100100.

In fact, we can be systematic about this and, remembering the order of the powers, we can use the ten powers to express all the numbers up to 1023 this way:

        0000000001 equals 1
        0000000010 equals 2
        0000000100 equals 4
        0000000101 equals 5
        0000000110 equals 6
        0000000111 equals 7, all the way up to

                    . . .
        1111111111 equals 1023.

Of course, we don't have to confine ourselves to ten powers of 2, we can have eleven powers, or fourteen, or fifty-three, or an infinite number. However it would get wearisome writing down an infinite number of 1's and 0's just to indicate whether each one of an infinite number of powers of 2 is used or is not used. So it is conventional to leave out all the high powers of 2 that are not used for a particular number and just begin with the highest power that *is* used and continue from there. In other words, leave out the unbroken line of zeros at the left. In that case, the numbers can be represented as:

$$1 \text{ equals } 1$$
$$10 \text{ equals } 2$$
$$11 \text{ equals } 3$$
$$100 \text{ equals } 4$$
$$101 \text{ equals } 5$$
$$110 \text{ equals } 6$$
$$111 \text{ equals } 7, \text{ and so on.}$$

Any number at all can be expressed by some combination of 1's and 0's in this fashion, and a few primitive tribes have actually used a number system like this. The first civilized mathematician to work it out systematically, however, was Gottfried Wilhelm Leibniz, about three centuries ago. He was amazed and gratified because he reasoned that 1, representing unity, was clearly a symbol for God, while 0 represented the nothingness which, aside from God, existed in the beginning. Therefore, if all numbers can be represented merely by the use of 1 and 0, surely this is the same as saying that God created the universe out of nothing.

Despite this awesome symbolism, this business of 1's and 0's made no impression whatsoever on practical men of affairs. It might be a fascinating mathematical curiosity, but no accountant is going to work with 1011011001 instead of 729.

But then it suddenly turned out that this two-based system of numbers (also called the "binary system," from the Latin word *binarius,* meaning "two at a time") is ideal for electronic computers.

After all, the two different digits, 1 and 0, can be matched in the computer by the two different positions of a particular switch: "on" and "off." Let "on" represent 1 and "off" represent 0. Then, if the machine contained ten switches, the number 1023 could be indicated as on-on-on-on-on-on-on-on-on-on; the number 729 could be on-off-on-on-off-on-on-off-off-on; and the number 100 could be off-off-off-on-on-off-off-on-off-off.

By adding more switches we can express any number we want simply by this on-off combination. It may seem complicated to us, but it is simplicity itself to the computer. In fact, no other conceivable system could be as simple—for the computer.

However, since we are only human beings, the question is, can *we* handle the two-based system? For instance, can we convert back and forth between two-based numbers and ordinary numbers? If we are shown 110001 in the two-based system, what does it mean in ordinary numbers?

Actually, this is not difficult. The two-based system uses powers of 2, starting at the extreme right with 2^0 and moving up a power at a time as we move leftward. So we can write 110001 with little numbers underneath to represent the exponents, thus $1\,1\,0\,0\,0\,1$. Only the exponents

$$5\,4\,3\,2\,1\,0$$

under the 1's are used, so 110001 represents 2^5 plus 2^4 plus 2^0 or 32 plus 16 plus 1. In other words, 110001 in the two-based system is 49 in ordinary numbers.

Working the other way is even simpler. You can, if you wish, try to fit the powers of 2 into an ordinary number by hit and miss, but you don't have to. There is a routine you can use which always works and I will describe it (though, if you will forgive me, I will not bother to explain *why* it works).

Suppose you wish to convert an ordinary number into the two-based system. You divide it by 2 and set the remainder to one side. (If the number is even, the remainder will be zero; if odd, it will be 1.) Working only with the whole-number portion of the quotient, you divide that by 2 again, and again set the remainder to one side and work only with the whole-number portion of the new quotient. When the whole-number portion of the quotient is reduced to 0 as a result of the repeated divisions by 2, you stop. The remainders, read backward, give the original number in the two-based system.

If this sounds complicated, it can be made simple enough by use of an example. Let's try 131:

131 divided by 2 is 65 with a remainder of 1
65 divided by 2 is 32 with a remainder of 1
32 divided by 2 is 16 with a remainder of 0
16 divided by 2 is 8 with a remainder of 0
8 divided by 2 is 4 with a remainder of 0
4 divided by 2 is 2 with a remainder of 0
2 divided by 2 is 1 with a remainder of 0
1 divided by 2 is 0 with a remainder of 1

In the two-based system, then, 131 is written 10000011.

With a little practice anyone who knows fourth-grade arithmetic can switch back and forth between ordinary numbers and two-based numbers.

The two-based system has the added value that it makes the ordinary operations of arithmetic childishly simple. In using ordinary numbers, we spend several years in the early grades memorizing the fact that 9 plus 5 is 14, that 8 times 3 is 24, and so on.

In two-based numbers, however, the only digits involved are 1 and 0, so there are only four possible sums of digits taken two at a time: 0 plus 0, 1 plus 0, 0 plus 1, and 1 plus 1. The first three are just what one would expect in ordinary arithmetic:

$$0 \text{ plus } 0 \text{ equals } 0$$
$$1 \text{ plus } 0 \text{ equals } 1$$
$$0 \text{ plus } 1 \text{ equals } 1$$

The fourth sum involves a slight difference. In ordinary arithmetic 1 plus 1 is 2, but there is no digit like 2 in the two-based system. There 2 is represented as 10. Therefore:

$$1 \text{ plus } 1 \text{ equals } 10 \text{ (put down 0 and carry 1)}$$

Imagine, then, how simple addition is in the two-based system. If you want to add 1001101 and 11001, the sum would look like this:

$$\begin{array}{r} 1001101 \\ 11001 \\ \hline 1100110 \end{array}$$

You can follow this easily from the addition table I've just given you, and by converting to ordinary numbers (as you ought also to be able to do) you will see that the addition is equivalent to 77 plus 25 equals 102.

It may seem to you that following the 1's and 0's is difficult indeed and that the ease of memorizing the rules of addition is more than made up for by the ease of losing track of the whole thing. This is true enough—for a human. In a computer, however, on-off switches are easily designed in such combinations as to make it possible for the on's and off's to follow the rules of addition in the two-based system. Computers don't get confused and surges of electrons bouncing this way and that add numbers by two-based addition in microseconds.

Of course (to get back to humans) if you want to add more than two numbers, you can always, at worst, break them up into groups of two. If you want to add 110, 101, 100, and 111, you can first add 110 and 101 to get 1011, then add 100 and 111 to get 1011, and finally add 1011 and 1011 to get 10110. (The last addition involves adding 1 plus 1 plus 1 as a result of carrying a 1 into a column which is already 1 plus 1. Well, 1 plus 1 is 10 and 10 plus 1 is 11, so 1 plus 1 plus 1 is 11, put down 1 and carry 1.)

Multiplication in the two-based system is even simpler. Again, there are only four possible combinations: 0 times 0, 0 times 1, 1 times 0, and 1 times 1. Here, each multiplication in the two-based system is exactly as it would be in ordinary numbers. In other words:

$$0 \text{ times } 0 \text{ is } 0$$
$$0 \text{ times } 1 \text{ is } 0$$
$$1 \text{ times } 0 \text{ is } 0$$
$$1 \text{ times } 1 \text{ is } 1$$

To multiply 101 by 1101, we would have

$$
\begin{array}{r}
101 \\
1101 \\
\hline
101 \\
000 \\
101 \\
101 \\
\hline
1000001 \\
\end{array}
$$

In ordinary numbers, this is equivalent to saying 5 times 13 is 65. Again, the computer can be designed to manipulate the on's and off's of its switches to match the requirements of the two-based multiplication table—and to do it with blinding speed.

It is possible to have a number system based on powers of 3, also (a three-based or "ternary" system). The series of numbers 3^0, 3^1, 3^2, 3^3, 3^4, and so on (that is, 1, 3, 9, 27, 81, and so on) can be used to express any finite number provided you are allowed to use up to two of each member of the series.

Thus 17 is 9 plus 3 plus 3 plus 1 plus 1, and 72 is 27 plus 27 plus 9 plus 9.

If you wanted to write the series of integers according to the three-based system, they would be: 1, 2, 10, 11, 12, 20, 21, 22, 100, 101, 102, 110, 111, 112, 120, 121, 122, 200, and so on.

You could have a four-based number system based on powers of 4, with each power used up to three times; a five-based number system based on powers of 5 with each power used up to four times; and so on.

To convert an ordinary number into any one of these other systems, you need only use a device similar to the one I have demonstrated for conversion into the two-based system. Where you repeatedly divide by 2 for the two-based system, you would repeatedly divide by 3 for the three-based system, by 4 for the four-based system, and so on.

Thus, I have already converted the ordinary number 131 into 11000001 by dividing 131 repeatedly by 2 and using the remainders. Suppose we divide 131 repeatedly by 3 instead and make use of the remainders:

131 divided by 3 is 43 with a remainder of 2
43 divided by 3 is 14 with a remainder of 1
14 divided by 3 is 4 with a remainder of 2
4 divided by 3 is 1 with a remainder of 1
1 divided by 3 is 0 with a remainder of 1

The number 131 in the three-based system, then, is made up of the remainders, working from the bottom up, and is 11212.

In similar fashion we can work out what 131 is in the four-based system, the five-based system, and so on. Here is a little table to give you the values of 131 up through the nine-based system:

two-based system	11000001
three-based system	11212
four-based system	2003

five-based system	1011
six-based system	335
seven-based system	245
eight-based system	203
nine-based system	155

You can check these by working through the powers. In the nine-based system, 155 is $1 \cdot 9^2$ plus $5 \cdot 9^1$ plus $5 \cdot 9^0$. Since 9^2 is 81, 9^1 is 9, and 9^0 is 1, we have 81 plus 45 plus 5, or 131. In the six-based system, 335 is $3 \cdot 6^2$ plus $3 \cdot 6^1$ plus $5 \cdot 6^0$. Since 6^2 is 36, 6^1 is 6, and 6^0 is 1, we have 108 plus 18 plus 5, or 131. In the four-based system, 2003 is $2 \cdot 4^3$ plus $0 \cdot 4^2$ plus $0 \cdot 4^1$ plus $3 \cdot 4^0$, and since 4^3 is 64, 4^2 is 16, 4^1 is 4, and 4^0 is 1, we have 128 plus 0 plus 0 plus 3, or 131.

The others you can work out for yourself if you choose.

But is there any point to stopping at a nine-based system? Can there be a ten-based system? Well, suppose we write 131 in the ten-based system by dividing it through by tens:

131 divided by 10 is 13 with a remainder of 1
13 divided by 10 is 1 with a remainder of 3
1 divided by 10 is 0 with a remainder of 1

And therefore 131 in the ten-based system is 131.

In other words, our ordinary numbers are simply the ten-based system, working on a series of powers of 10: 10^0, 10^1, 10^2, 10^3, and so on. The number 131 is equal to $1 \cdot 10^2$ plus $3 \cdot 10^1$ plus $1 \cdot 10^0$. Since 10^2 is 100, 10^1 is 10, and 10^0 is 1, this means we have 100 plus 30 plus 1, or 131.

There is nothing basic or fundamental about ordinary numbers then. They are based on the powers of 10 because we have ten fingers and counted on our fingers to begin with, but the powers of any other number will fulfill all the mathematical requirements.

Thus we can go on to an eleven-based system and a twelve-based system. Here, one difficulty arises. The number of digits (counting zero) that is required for any system is equal to the number used as base.

In the two-based system, we need two different digits, 0 and 1. In the three-based system, we need three different digits, 0, 1, and 2. In the familiar ten-based system, we need, of course, ten different digits, 0, 1, 2, 3, 4, 5, 6, 7, 8, and 9.

It follows, then, that in the eleven-based system we will need eleven different digits and in the twelve-based system twelve different digits. Let's write @ for the eleventh digit and # for the twelfth. In ordinary ten-based numbers, @ is 10 and # is 11.

Thus, 131 in the eleven-based system is:

131 divided by 11 is 11 with a remainder of 10 (@)
11 divided by 11 is 1 with a remainder of 0
1 divided by 11 is 0 with a remainder of 1

so that 131 in the eleven-based system is 10 @.
And in the twelve-based system:

131 divided by 12 is 10 with a remainder of 11 (#)
10 divided by 12 is 0 with a remainder of 10 (@)

so that 131 in the twelve-based system is @#.

And we can go up and up and up and have a 4583-based system if we wanted (but with 4583 different digits, counting the zero).

Now all the number systems may be valid, but which system is most convenient? As one goes to higher and higher bases, numbers become shorter and shorter. Though 131 is 11000001 in the two-based system, it is 131 in the ten-based system and @# in the twelve-based system. It moves from eight digits to three digits to two digits. In fact, in a 131-based system (and higher) it would be down to a single digit. In a way, this represents increasing convenience. Who needs long numbers?

However, the number of different digits used in constructing numbers goes up with the base and this is an increasing inconvenience. Somewhere there is an intermediate base in which the number of different digits isn't too high and the number of digits in the usual numbers we use isn't too great.

Naturally it would seem to us that the ten-based system is just right. Ten different digits to memorize doesn't seem

33

too high a price to pay for using only four digit combinations to make up any number under ten thousand.

Yet the twelve-based system has been touted now and then. Four digit combinations in the twelve-based system will carry one up to a little over twenty thousand, but that seems scarcely sufficient recompense for the task of learning to manipulate two extra digits. (School children would have to learn such operations as @ plus 5 is 13 and # times 4 is 38.)

But here another point arises. When you deal with any number system, you tend to talk in round numbers: 10, 100, 1000, and so on. Well, 10 in the ten-based system is evenly divisible by 2 and 5 and that is all. On the other hand, 10 in the twelve-based system (which is equivalent to 12 in the ten-based system) is evenly divisible by 2, 3, 4, and 6. This means that a twelve-based system would be more adaptable to commercial transactions and, indeed, the twelve-based system is used every time things are sold in dozens (12's) and grosses (144's) for 12 is 10 and 144 is 100 in the twelve-based system.

In this age of computers, however, the attraction is toward a two-based system. And while a two-based system is an uncomfortable and unaesthetic mélange of 1's and 0's, there is a compromise possible.

A two-based system is closely related to an eight-based system, for 1000 on the two-based system is equal to 10 on the eight-based system, or, if you'd rather, 2^3 equals 8^1. We could therefore set up a correspondence as follows:

Two-based System	Eight-based System
000	0
001	1
010	2
011	3
100	4
101	5
110	6
111	7

This would take care of *all* the digits (including zero) in the eight-based system and *all* the three-digit combinations (including 000) in the two-based system.

Therefore any two-based number could be broken up into groups of three digits (with zeros added to the left if necessary) and converted into an eight-based number by using the table I've just given you. Thus, the two-based number 1110010000101001110 could be broken up as 111,001,000,010,100,110 and written as the eight-based number, 710246. On the other hand, the eight-based number 33574 can be written as the two-based number 011011101111100 almost as fast as one can write, once one learns the table.

In other words, if we switched from a ten-based system to an eight-based system, there would be a much greater understanding between ourselves and our machines and who knows how much faster science would progress.

Of course, such a switch isn't practical, but just think— Suppose that, originally, primitive man had learned to count on his eight fingers only and had left out those two awkward and troublesome thumbs.

3. Varieties of the Infinite

THERE ARE a number of words that publishers like to get into the titles of science-fiction books as an instant advertisement to possible fans casually glancing over a display that these books are indeed science fiction. Two such words are, of course, *space* and *time*. Others are *Earth* (capitalized), *Mars, Venus, Alpha Centauri, tomorrow, stars, sun, asteroids,* and so on. And one—to get to the nub of this chapter—is *infinity*.

One of the best s.f. titles ever invented, in my opinion, is John Campbell's *Invaders from the Infinite*. The word *invaders* is redolent of aggression, action, and suspense, while *infinite* brings up the vastness and terror of outer space.

Donald Day's indispensable *Index to the Science Fiction Magazines* lists "Infinite Brain," "Infinite Enemy," "Infinite

Eye," "Infinite Invasion," "Infinite Moment," "Infinite Vision," and "Infinity Zero" in its title index, and I am sure there are many other titles containing the word.

Yet with all this exposure, with all this familiar use, do we know what *infinite* and *infinity* mean? Perhaps not all of us do.

We might begin, I imagine, by supposing that infinity was a large number; a very large number; in fact, the largest number that could exist.

If so, that would at once be wrong, for infinity is not a large number or any kind of number at all; at least of the sort we think of when we say "number." It certainly isn't the largest number that could exist, for there isn't any such thing.

Let's sneak up on infinity by supposing first that you wanted to write out instructions to a bright youngster, telling him how to go about counting the 538 people who had paid to attend a lecture. There would be one particular door through which all the audience would leave in single file. The youngster need merely apply to each person one of the various integers in the proper order: 1, 2, 3, and so on.

The phrase "and so on" implies continuing to count until all the people have left, and the last person who leaves has received the integer 538. If you want to make the order explicit, you might tell the boy to count in the following fashion and then painstakingly list all the integers from 1 to 538. This would undoubtedly be unbearably tedious, but the boy you are dealing with is bright and knows the meaning of a gap containing a dotted line, so you write: "Count thus: 1, 2, 3, . . . , 536, 537, 538." The boy will then understand (or should understand) that the dotted line indicates a gap to be filled by all the integers from 4 to 535 inclusive, in order and without omission.

Suppose you didn't know what the number of the audience was. It might be 538 or 427 or 651. You could instruct the boy to count until an integer had been given to the last man, whatever the man, whatever the integer. To express that symbolically, you could write thus: "Count: 1, 2, 3, . . . , $n - 2$, $n - 1$, n." The bright boy would understand that *n* routinely represents some unknown but definite integer.

Now suppose the next task you set your bright youngster was to count the number of men entering a door, filing through a room, out a second door, around the building, and through the first door again, the men forming a continuous closed system.

Imagine both marching men and counting boy to be completely tireless and willing to spend an eternity in their activities. Obviously the task would be endless. There would be no last man at all, ever, and there is no last integer at all, ever. (Any integer, however large, even if it consisted of a series of digits stretching in microscopic size from here to the farthest star, can easily be increased by 1.)

How do we write instructions for the precise counting involved in such a task. We can write: "Count thus: 1, 2, 3, and so on endlessly."

The phrase "and so on endlessly" can be written in shorthand, thus, ∞.

The statement "1, 2, 3, . . . , ∞" should be read "one, two, three, and so on endlessly" or "one, two, three, and so on without limit," but it is usually read, "one, two, three, and so on to infinity." Even mathematicians introduce infinity here, and George Gamow, for instance, has written a most entertaining book entitled just that: *One, Two, Three . . . Infinity*.

It might seem that using the word *infinity* is all right, since it comes from a Latin word meaning "endless," but nevertheless it would be better if the Anglo-Saxon were used in this case. The phrase "and so on endlessly" can't be mistaken. Its meaning is clear. The phrase "and so on to infinity," on the other hand, inevitably gives rise to the notion that infinity is some definite, though very huge, integer and that once we reach it we can stop.

So let's be blunt. Infinity is not an integer or any number of a kind with which we are familiar. It is a quality; a quality of endlessness. And any set of objects (numbers or otherwise) that is endless can be spoken of as an "infinite series" or an "infinite set." The list of integers from 1 on upward is an example of an "infinite set."

Even though ∞ is not a number, we can still put it through certain arithmetical operations. We can do that much for any symbol. We can do it for letters in algebra and write $a + b = c$. Or we can do it for chemical formulas and write: $CH_4 + 3O_2 = CO_2 + 2H_2O$. Or we can do it for abstractions, such as: Man + Woman = Trouble.

The only thing we must remember is that in putting symbols that are not integers through arithmetical paces, we ought not to be surprised if they don't follow the ordinary rules of arithmetic which, after all, were originally worked out to apply specifically to integers.

For instance, $3 - 2 = 1$, $17 - 2 = 15$, $4875 - 2 = 4873$. In general, any integer, once 2 is subtracted, becomes a different integer. Anything else is unthinkable.

But now suppose we subtract 2 from the unending series of integers. For convenience sake, we can omit the first two integers, 1 and 2, and start the series: 3, 4, 5, and so on endlessly. You see, don't you, that you can be just as endless starting the integers at 3 as at 1, so that you can write: $3, 4, 5, \ldots, \infty$.

In other words, when two items are subtracted from an infinite set, what remains is still an infinite set. In symbols, we can write this: $\infty - 2 = \infty$. This looks odd because we are used to integers, where subtracting 2 makes a difference. But infinity is not an integer and works by different rules. (This can't be repeated often enough.)

For that matter, if you lop off the first 3 integers or the first 25 or the first 1000000000000, what is left of the series of integers is still endless. You can always start, say, with 1000000000001, 1000000000002, and go on endlessly. So $\infty - n = \infty$, where n represents any integer, however great.

In fact, we can be more startling than that. Suppose we consider only the even integers. We would have a series that would go: 2, 4, 6, and so on endlessly. It would be an infinite series and could therefore be written: $2, 4, 6, \ldots, \infty$. In the same way, the odd integers would form an infinite series and could be written: $1, 3, 5, \ldots, \infty$.

Now, then, suppose you went through the series of integers and crossed out every even integer you came to, thus: 1, 2̸, 3, 4̸, 5, 6̸, 7, 8̸, 9, 1̸0̸, 11, 1̸2̸, $\ldots, \infty$. From the infinite series of integers you would have eliminated an infinite series of even integers and you would have left behind an infinite series of odd integers. This can be symbolized as $\infty - \infty = \infty$.

Furthermore, it could work the other way about. If you started with the even integers only and added one odd integer, or two, or five, or a trillion, you would still merely have an unending series, so that $\infty + n = \infty$. In fact, if you added the unending series of odd integers to the unending series of even integers, you would simply have the unending series of all integers, or: $\infty + \infty = \infty$.

By this point, however, it is just possible that some of you may suspect me of pulling a fast one.

After all, in the first 10 integers, there are 5 even integers and 5 odd ones; in the first 1000 integers, there are 500 even integers and 500 odd integers; and so on. No matter how many consecutive integers we take, half are always even and half are odd.

Therefore, although the series 2, 4, 6, . . . *is* endless, the total can only be half as great as the total of the also endless series 1, 2, 3, 4, 5, 6. . . . And the same is true for the series 1, 3, 5, . . . , which, though endless, is only half as great as the series of all integers.

And so (you might think) in subtracting the set of even integers from the set of all integers to obtain the set of odd integers, what we are doing can be represented as: $\infty - \frac{1}{2}\infty = \frac{1}{2}\infty$. That, you might think with a certain satisfaction, "makes sense."

To answer that objection, let's go back to counting the unknown audience at the lecture. Our bright boy, who has been doing all our counting, and is tired of it, turns to you and asks, "How many seats are there in the lecture hall?" You answer, "640."

He thinks a little and says, "Well, I see that every seat is taken. There are no empty seats and there is no one standing."

You, having equally good eyesight, say, "That's right."

"Well, then," says the boy, "why count them as they leave. We know right now that there are exactly 640 spectators."

And he's correct. If two series of objects (*A* series and *B* series) just match up so that there is one and only one *A* for every *B* and one and only one *B* for every *A*, then we know that the total number of *A* objects is just equal to the total number of *B* objects.

In fact, this is what we do when we count. If we want

39

to know how many teeth there are in the fully equipped human mouth, we assign to each tooth one and only one number (in order) and we apply each number to one and only one tooth. (This is called placing two series into "one-to-one correspondence.") We find that we need only 32 numbers to do this, so that the series 1, 2, 3, . . . , 30, 31, 32 can be exactly matched with the series one tooth, next tooth, next tooth, . . . , next tooth, next tooth, last tooth.

And therefore, we say, the number of teeth in the fully-equipped human mouth is the same as the number of integers from 1 to 32 inclusive. Or, to put it tersely and succinctly: there are 32 teeth.

Now we can do the same for the set of even integers. We can write down the even integers and give each one a number. Of course, we can't write down some and get started anyway. We can write the number assigned to each even integer directly above it, with a double-headed arrow, so:

$$1 \quad 2 \quad 3 \quad 4 \quad 5 \quad 6 \quad 7 \quad 8 \quad 9 \quad 10 \ldots$$
$$\updownarrow \quad \updownarrow \quad \updownarrow \quad \updownarrow \quad \updownarrow \quad \updownarrow \quad \updownarrow \quad \updownarrow \quad \updownarrow \quad \updownarrow$$
$$2 \quad 4 \quad 6 \quad 8 \quad 10 \quad 12 \quad 14 \quad 16 \quad 18 \quad 20 \ldots$$

We can already see a system here. Every even integer is assigned one particular number and no other, and you can tell what the particular number is by dividing the even integer by 2. Thus, the even integer 38 has the number 19 assigned to it and no other. The even integer 24618 has the number 12309 assigned to it. In the same way, any given number in the series of all integers can be assigned to one and only one even integer. The number 538 is applied to even integer 1076 and to no other. The number 29999999 is applied to even integer 59999998 and no other; and so on.

Since every number in the series of even integers can be applied to one and only one number in the series of all integers and vice versa, the two series are in one-to-one correspondence and are equal. The number of even integers then is equal to the number of all integers. By a similar argument, the number of odd integers is equal to the number of all integers.

You may object by saying that when all the even integers (or odd integers) are used up, there will still be fully half the series of all integers left over. Maybe so, but this argument has no meaning since the series of even integers (or odd integers) will never be used up.

Therefore, when we say that "all integers" minus "even integers" equals "odd integers," this *is* like saying $\infty - \infty = \infty$, and terms like $\frac{1}{2}\infty$ can be thrown out.

In fact, in subtracting even integers from all integers, we are crossing out every other number and thus, in a way, dividing the series by 2. Since the series is still unending, $\infty/2 = \infty$ anyway, so what price half of infinity?

Better yet, if we crossed out every other integer in the series of even integers, we would have an unending series of integers divisible by 4; and if we crossed out every other integer in that series, we would have an unending series of integers divisible by 8, and so on endlessly. Each one of these "smaller" series could be matched up with the series of all integers in one-to-one correspondence. If an unending series of integers can be divided by 2 endlessly, and still remain endless, then we are saying that $\infty/\infty = \infty$.

If you doubt that endless series that have been drastically thinned out can be put into one-to-one correspondence with the series of all integers, just consider those integers that are multiples of one trillion. You have: 1,000,000,-000,000; 2,000,000,000,000; 3,000,000,000,000; . . . ; ∞. These are matched up with 1, 2, 3, . . . , ∞. For any number in the set of "trillion-integers," say 4,856,000,000,-000,000, there is one and only one number in the set of all integers, which, in this case, is 4856. For any number in the set of all integers, say 342, there is one and only one number in the set of "trillion-integers," in this case, 342,000,000,000,000. Therefore, there are as many integers divisible by a trillion as there are integers altogether.

It works the other way around, too. If you place between each number the midway fraction, thus: $\frac{1}{2}$, 1, $1\frac{1}{2}$, 2, $2\frac{1}{2}$, 3, $3\frac{1}{2}$, . . . , ∞, you are, in effect, doubling the number of items in the series and yet this new series can be put into one-to-one correspondence with the set of integers, so that $2\infty = \infty$. In fact, if you keep on doing it indefinitely, putting in all the fourths, then all the eighths, then all the sixteenths, you can still keep the resulting series in

41

one-to-one correspondence with the set of all integers so that $\infty \cdot \infty = \infty^2 = \infty$.

This may seem too much to swallow. How can all the fractions be lined up so that we can be sure that each one is getting one and only one number? It is easy to line up integers, 1, 2, 3, or even integers, 2, 4, 6, or even prime numbers 2, 3, 5, 7, 11. . . . But how can you line up fractions and be sure that all are included, even fancy ones like $\frac{14899}{2725523}$ and $\frac{689444473}{2}$.

There are, however, several ways to make up an inclusive list of fractions. Suppose we first list all the fractions in which the numerator and denominator add up to 2. There is only one of these: $\frac{1}{1}$. Then list those fractions where the numerator and denominator add up to 3. There are two of these: $\frac{2}{1}$ and $\frac{1}{2}$. Then we have $\frac{3}{1}$, $\frac{2}{2}$, and $\frac{1}{3}$, where the numerator and denominator add up to 4. Then we have $\frac{4}{1}$, $\frac{3}{2}$, $\frac{2}{3}$, and $\frac{1}{4}$. In each group, you see, we place the fractions in order of decreasing numerator and increasing denominator.

If we make such a list: $\frac{1}{1}$, $\frac{2}{1}$, $\frac{1}{2}$, $\frac{3}{1}$, $\frac{2}{2}$, $\frac{1}{3}$, $\frac{4}{1}$, $\frac{3}{2}$, $\frac{2}{3}$, $\frac{1}{4}$, $\frac{5}{1}$, $\frac{4}{2}$, $\frac{3}{3}$, $\frac{2}{4}$, $\frac{1}{5}$ and so on endlessly, we can be assured that any particular fraction, no matter how complicated, will be included if we proceed far enough. The fraction $\frac{14899}{2725523}$ will be in that group of fractions in which the numerator and denominator add up to 2740422, and it will be the 2725523rd of the group. Similarly, $\frac{689444473}{2}$ will be the second fraction in the group in which the numerator and the denominator add up to 689444475. Every possible fraction will thus have its particular assigned place in the series.

It follows, then, that every fraction has its own number and that no fraction will be left out. Moreover, every number has its own fraction and no number is left out. The series of all fractions is put into a one-to-one correspondence with the series of all integers, and thus the number of all fractions is equal to the number of all integers.

(In the list of fractions above, you will see that some are equal in value. Thus, $\frac{1}{2}$ and $\frac{2}{4}$ are listed as different fractions, but both have the same value. Fractions like $\frac{1}{1}$, $\frac{2}{2}$, and $\frac{3}{3}$ not only have the same value but that value is that of an integer, 1. All this is all right. It shows that the total number of fractions is equal to the total number of integers even though in the series of fractions, the value of each particular fraction, and all integral values as well, is repeated many times; in fact, endlessly.)

By now you may have more or less reluctantly decided that all unendingness is the same unendingness and that "infinity" is "infinity" no matter what you do to it.

Not so!

Consider the points in a line. A line can be marked off at equal intervals, and the marks can represent points which are numbered 1, 2, 3, and so on endlessly, if you imagine the line continuing endlessly. The midpoints between the integer-points can be marked ½, 1½. 2½, . . . , and then the thirds can be marked and the fourths and the fifths and indeed all the unending number of fractions can be assigned to some particular point.

It would seem then that every point in the line would have some fraction or other assigned to it. Surely there would be no point in the line left out after an unending number of fractions had been assigned to it?

Oh, wouldn't there?

There is a point on the line, you see, that would be represented by a value equal to the square root of two ($\sqrt{2}$). This can be shown as follows. If you construct a square on the line with each side exactly equal to the interval of one integer already marked off on the line, then the diagonal of the square would be just equal to $\sqrt{2}$. If that diagonal is laid down on the line, starting from the zero point, the end of that diagonal coincides with the point on the line which can be set equal to $\sqrt{2}$.

Now the catch is that the value of $\sqrt{2}$ cannot be represented by a fraction; by any fraction; by any conceivable fraction. This was proved by the ancient Greeks and the proof is simple but I'll ask you to take my word for it here to save room. Well, if all the fractions are assigned to various points in the line, at least one point, that which corresponds to $\sqrt{2}$, will be left out.

All numbers which can be represented as fractions are "rational numbers" because a fraction is really the ratio of two numbers, the numerator and the denominator. Numbers which cannot be represented as fractions are "irrational numbers" and $\sqrt{2}$ is by no means the only one of those, although it was the first such to be discovered. Most square roots, cube roots, fourth roots, etc., are irrationals, so are most sines, cosines, tangents, etc., so are numbers involving pi (π), so are logarithms.

In fact, the set of irrational numbers is unending. It can be shown that between any two points represented by rational numbers on a line, however close those two points are, there is always at least one point represented by an irrational number.

Together, the rational numbers and irrational numbers are spoken of as "real numbers." It can be shown that any given real numbers can be made to correspond to one and only one point in a given line; and that any point in the line can be made to correspond to one and only one real number. In other words, a point in a line which can't be assigned a fraction, can always be assigned an irrational. No point can be missed by both categories.

The series of real numbers and the series of points in a line are therefore in one-to-one correspondence and are equal.

Now the next question is: Can the series of all real numbers, or of all points in a line (the two being equivalent), be set into a one-to-one correspondence with the series of integers. The answer is, *No!*

It can be shown that no matter how you arrange your real numbers or your points, no matter what conceivable system you use, an endless number of either real numbers or points will always be left out. The result is that we are in the same situation as that in which we are faced with an audience in which all seats are taken and there are people standing. We are forced to conclude that there are more people than seats. And so, in the same way, we are forced to conclude that there are more real numbers, or points in a line, than there are integers.

If we want to express the endless series of points by symbols, we don't want to use the symbol ∞ for "and so on endlessly," since this has been all tied up with integers and rational numbers generally. Instead, the symbol C is usually used, standing for *continuum,* since all the points in a line represent a continuous line.

We can therefore write the series: Point 1, Point 2, Point 3, ..., C.

Now we have a variety of endlessness that is different

and *more intensely endless* than the endlessness represented by "ordinary infinity."

This new and more intense endlessness also has its peculiar arithmetic. For instance, the points in a short line can be matched up one-for-one with the points in a long line, or the points in a plane, or the points in a solid. In fact, let's not prolong the agony, and say at once that there are as many points in a line a millionth of an inch long as there are points in all of space.

About 1895 the German mathematician Georg Cantor worked out the arithmetic of infinity and also set up a whole series of different varieties of endlessness, which he called "transfinite numbers."

He represented these transfinite numbers by the letter *aleph*, which is the first letter of the Hebrew alphabet and which looks like this: $\aleph$

The various transfinites can be listed in increasing size or, rather, in increasing intensity of endlessness by giving each one a subscript, beginning with zero. The very lowest transfinite would be "aleph-null," then there would be "aleph-one," "aleph-two," and so on, endlessly.

This could be symbolized as: $\aleph_0$, $\aleph_1$, $\aleph_2$, ..., $\aleph_\infty$

Generally, whatever you do to a particular transfinite number in the way of adding, subtracting, multiplying, or dividing, leaves it unchanged. A change comes only when you raise a transfinite to a transfinite power equal to itself (not to a transfinite power less than itself). Then it is increased to the next higher transfinite. Thus:

$$\aleph_0^{\aleph_0} = \aleph_1; \quad \aleph_1^{\aleph_1} = \aleph_2; \text{ and so on.}$$

What we usually consider as infinity, the endlessness of the integers, has been shown to be equal to aleph-null. In other words: $\infty = \aleph_0$. And so the tremendous vastness of ordinary infinity turns out to be the very smallest of all the transfinites.

That variety of endlessness which we have symbolized as *C may* be represented by aleph-one so that $C = \aleph_1$, but this has not been proved. No mathematician has yet been able to prove that there is any infinite series which has an endlessness more intense than the endlessness of the integers but less intense than the endlessness of the points in a line. However, neither has any mathematician been able to prove that such an intermediate endlessness does *not* exist.

If the continuum *is* equal to aleph-one, then we can finally write an equation for our friend "ordinary infinity" whch will change it:

$$\infty^{\infty} = C.$$

Finally, it has been shown that the endlessness of all the curves that can be drawn on a plane is even more intense than the endlessness of points in a line. In other words, there is no way of lining up the curves so that they can be matched one-to-one with the points in a line, without leaving out an unending series of the curves. This endlessness of curves may be equal to aleph-two, but that hasn't been proved yet, either.

And that is all. Assuming that the endlessness of integers is aleph-null, and the endlessness of points is aleph-one, and the endlessness of curves is aleph-two, we have come to the end. Nobody has ever suggested any variety of endlessness which could correspond to aleph-three (let alone to aleph-thirty or aleph-three-million).

As John E. Freund says in his book *A Modern Introduction to Mathematics*[1] (a book I recommend to all who found this article in the least interesting), "It seems that our imagination does not permit us to count beyond *three* when dealing with infinite sets."

Still, if we now return to the title *Invaders from the Infinite*, I think we are entitled to ask, with an air of phlegmatic calm, "Which infinite? Just aleph-null? Nothing more?"

[1] New York: Prentice-Hall, 1956.

4. A Piece of Pi

IN MY ESSAY "Those Crazy Ideas," which appeared in *Fact and Fancy*, I casually threw in a footnote to the effect that $e^{\pi i} = -1$. Behold, a good proportion of the comment which I received thereafter dealt not with the essay itself but with that footnote (one reader, more in sorrow than in anger, proved the equality, which I had neglected to do).

My conclusion is that some readers are interested in these odd symbols. Since I am, too (albeit I am not really a mathematician, or anything else), the impulse is irresistible to pick up one of them, say π, and talk about it in this chapter and the next. In Chapter 6, I will discuss i.

In the first place, what is π? Well, it is the Greek letter *pi* and it represents the ratio of the length of the perimeter of a circle to the length of its diameter. *Perimeter* is from the Greek *perimetron*, meaning "the measurement around," and *diameter* from the Greek *diametron*, meaning "the measurement through." For some obscure reason, while it is customary to use perimeter in case of polygons, it is also customary to switch to the Latin *circumference* in speaking of circles. This is all right, I suppose (I am no purist) but it obscures the reason for the symbol π.

Back about 1600 the English mathematician William Oughtred, in discussing the ratio of a circle's perimeter to its diameter, used the Greek letter π to symbolize the perimeter and the Greek letter δ (delta) to symbolize the diameter. They were the first letters, respectively, of *perimetron* and *diametron*.

Now mathematicians often simplify matters by setting values equal to unity whenever they can. For instance, they might talk of a circle of unit diameter. In such a circle, the length of the perimeter is numerically equal to the ratio of perimeter to diameter. (This is obvious to some of you,

47

I suppose, and the rest of you can take my word for it.) Since in a circle of unit diameter the perimeter equals the ratio, the ratio can be symbolized by π, the symbol of the perimeter. And since circles of unit diameter are very frequently dealt with, the habit becomes quickly ingrained.

The first top-flight man to use π as the symbol for the ratio of the length of a circle's perimeter to the length of its diameter was the Swiss mathematician Leonhard Euler, in 1737, and what was good enough for Euler was good enough for everyone else.

Now I can go back to calling the distance around a circle the circumference.

But what *is* the ratio of the circumference of a circle to its diameter in actual numbers?

This apparently is a question that always concerned the ancients even long before pure mathematics was invented. In any kind of construction past the hen-coop stage you must calculate in advance all sorts of measurements, if you are not perpetually to be calling out to some underling, "You nut, these beams are all half a foot too short." In order to make the measurements, the universe being what it is, you are forever having to use the value of π in multiplications. Even when you're not dealing with circles, but only with angles (and you can't avoid angles) you will bump into π.

Presumably, the first empirical calculators who realized that the ratio was important, determined the ratio by drawing a circle and actually measuring the length of the diameter and the circumference. Of course, measuring the length of the circumference is a tricky problem that can't be handled by the usual wooden foot-rule, which is far too inflexible for the purpose.

What the pyramid-builders and their predecessors probably did was to lay a linen cord along the circumference very carefully, make a little mark at the point where the circumference was completed, then straighten the line and measure it with the equivalent of a wooden foot-rule. (Modern theoretical mathematicians frown at this and make haughty remarks such as "But you are making the unwarranted assumption that the line is the same length when it is straight as when it was curved." I imagine the honest workman organizing the construction of the local temple, faced with such an objection, would have solved matters by throwing the objector into the river Nile.)

Anyway, by drawing circles of different size and making enough measurements, it undoubtedly dawned upon architects and artisans, very early in the game, that the ratio was always the same in all circles. In other words, if one circle had a diameter twice as long or 1⅝ as long as the diameter of a second, it would also have a circumference twice as long or 1⅝ as long. The problem boiled down, then, to finding not the ratio of the particular circle you were interested in using, but a universal ratio that would hold for all circles for all time. Once someone had the value of π in his head, he would never have to determine the ratio again for any circle.

As to the actual value of the ratio, as determined by measurement, that depended, in ancient times, on the care taken by the person making the measurement and on the value he placed on accuracy in the abstract. The ancient Hebrews, for instance, were not much in the way of construction engineers, and when the time came for them to build their one important building (Solomon's temple), they had to call in a Phoenician architect.

It is to be expected, then, that the Hebrews in describing the temple would use round figures only, seeing no point in stupid and troublesome fractions, and refusing to be bothered with such petty and niggling matters when the House of God was in question.

Thus, in Chapter 4 of 2 Chronicles, they describe a "molten sea" which was included in the temple and which was, presumably, some sort of container in circular form. The beginning of the description is in the second verse of that chapter and reads: "Also he made a molten sea of ten cubits from brim to brim, round in compass, and five cubits the height thereof; and a line of thirty cubits did compass it round about."

The Hebrews, you see, did not realize that in giving the diameter of a circle (as ten cubits or as anything else) they automatically gave the circumference as well. They felt it necessary to specify the circumference as thirty cubits and in so doing revealed the fact that they considered π to be equal to exactly 3.

There is always the danger that some individual, too wedded to the literal words of the Bible, may consider 3 to be the divinely ordained value of π in consequence. I wonder if this may not have been the motive of the simple soul in some state legislature who, some years back, introduced a bill which would have made π legally equal to 3 inside the bounds of the state. Fortunately, the bill did not pass or all the wheels in that state (which would, of course, have respected the laws of the state's august legislators) would have turned hexagonal.

In any case, those ancients who were architecturally sophisticated knew well, from their measurements, that the value of π was distinctly more than 3. The best value they had was $\frac{22}{7}$ (or $3\frac{1}{7}$, if you prefer) which really isn't bad and is still used to this day for quick approximations. Decimally, $\frac{22}{7}$ is equal, roughly, to 3.142857 . . . , while π is equal, roughly, to 3.141592. . . . Thus, $\frac{22}{7}$ is high by only 0.04 percent or 1 part in 2500. Good enough for most rule-of-thumb purposes.

Then along came the Greeks and developed a system of geometry that would have none of this vile lay-down-a-string-and-measure-it-with-a-ruler business. That, obviously, gave values that were only as good as the ruler and the string and the human eye, all of which were dreadfully imperfect. Instead, the Greeks went about deducing what the value of π must be once the perfect lines and curves of the ideal plane geometry they had invented were taken properly into account.

Archimedes of Syracuse, for instance, used the "method of exhaustion" (a forerunner of integral calculus, which Archimedes might have invented two thousand years before Newton if some kind benefactor of later centuries had only sent him the Arabic numerals via a time machine) to calculate π.

To get the idea, imagine an equilateral triangle with its vertexes on the circumference of a circle of unit diameter. Ordinary geometry suffices to calculate exactly the perimeter of that triangle. It comes out to $3\sqrt{3/2}$, if you are curious, or 2.598076. . . . This perimeter has to be less than that of the circle (that is, than the value of π), again by elementary geometrical reasoning.

Next, imagine the arcs between the vertexes of the triangle divided in two so that a regular hexagon (a six-sided figure) can be inscribed in the circle. Its perimeter can be determined also (it is exactly 3) and this can be shown to be larger than that of the triangle but still less than that of the circle. By proceeding to do this over and over again, a regular polygon with 12, 24, 48 . . . sides can be inscribed.

The space between the polygon and the boundary of the circle is steadily decreased or "exhausted" and the polygon approaches as close to the circle as you wish, though it never really reaches it. You can do the same with a series of equilateral polygons that circumscribe the circle (that lay outside it, that is, with their sides tangent to the circle) and get a series of decreasing values that approach the circumference of the circle.

In essence, Archimedes trapped the circumference between a series of numbers that approached π from below, and another that approached it from above. In this way π could be determined with any degree of exactness, provided you were patient enough to bear the tedium of working with polygons of large numbers of sides.

Archimedes found the time and patience to work with polygons of ninety-six sides and was able to show that the value of π was a little below $\frac{22}{7}$ and a little above the slightly smaller fraction $\frac{223}{71}$.

Now the average of these two fractions is $\frac{3123}{994}$ and the decimal equivalent of that is 3.141851. . . . This is more than the true value of π by only 0.0082 percent or 1 part in 12,500.

Nothing better than this was obtained, in Europe, at least, until the sixteenth century. It was then that the fraction $\frac{355}{113}$ was first used as an approximation of π. This is really the best approximation of π that can be expressed as a reasonably simple fraction. The decimal value of $\frac{355}{113}$ is 3.14159292 . . . , while the true value of π is 3.14159265. . . . You can see from that that $\frac{355}{113}$ is higher than the true value by only 0.000008 percent, or by one part in 12,500,000.

Just to give you an idea of how good an approximation $\frac{355}{113}$ is, let's suppose that the earth were a perfect sphere with a diameter of exactly 8000 miles. We could then calculate the length of the equator by multiplying 8000 by π. Using the approximation $\frac{355}{113}$ for π, the answer comes out 25,132.7433 . . . miles. The true value of π would give the answer 25,132.7412 . . . miles. The difference would come to about 11 feet. A difference of 11 feet in calculating the circumference of the earth might well be reckoned as negligible. Even the artificial satellites that have brought our geography to new heights of precision haven't supplied us with measurements within that range of accuracy.

It follows then that for anyone but mathematicians, $\frac{355}{113}$ is as close to π as it is necessary to get under any but the most unusual circumstances. And yet mathematicians have their own point of view. They can't be happy without the true value. As far as they are concerned, a miss, however close, is as bad as a megaparsec.

The key step toward the true value was taken by François Vieta, a French mathematician of the sixteenth century. He is considered the father of algebra because, among other things, he introduced the use of letter symbols for unknowns, the famous x's and y's, which most of us have had to, at one time or another in our lives, face with trepidation and uncertainty.

Vieta performed the algebraic equivalent of Archimedes' geometric method of exhaustion. That is, instead of setting up an infinite series of polygons that came closer and closer to a circle, he deduced an infinite series of fractions which could be evaluated to give a figure for π. The greater the number of terms used in the evaluation, the closer you were to the true value of π.

I won't give you Vieta's series here because it involves square roots and the square roots of square roots and the square roots of square roots of square roots. There is no point in involving one's self in that when other mathematicians derived other series of terms (always an infinite series) for the evaluation of π; series much easier to write.

For instance, in 1673 the German mathematician Gottfried Wilhelm von Leibniz (who first worked out the binary system—see Chapter 2) derived a series which can be expressed as follows:

$$\pi = \tfrac{4}{1} - \tfrac{4}{3} + \tfrac{4}{5} - \tfrac{4}{7} + \tfrac{4}{9} - \tfrac{4}{11} + \tfrac{4}{13} - \tfrac{4}{15} \ldots$$

Being a naïve nonmathematician myself, with virtually no mathematical insight worth mentioning, I thought, when I first decided to write this essay, that I would use the Leibniz series to dash off a short calculation and show you how it would give π easily to a dozen places or so. However, shortly after beginning, I quit.

You may scorn my lack of perseverance, but any of you are welcome to evaluate the Leibniz series just as far as it is written above, to $\frac{4}{15}$, that is. You can even drop me a postcard and tell me the result. If, when you finish, you are disappointed to find that your answer isn't as close to π as the value of $\frac{355}{113}$, don't give up. Just add more terms. Add $\frac{4}{17}$ to your answer, then subtract $\frac{4}{19}$, then add $\frac{4}{21}$ and subtract $\frac{4}{23}$, and so on. You can go on as long as you want to, and if any of you find out how many terms it takes to improve on $\frac{355}{113}$, drop me a line and tell me that, too.

Of course, all this may disappoint you. To be sure, the endless series *is* a mathematical representation of the true and exact value of π. To a mathematician, it is as valid a way as any to express that value. But if you want it in the form of an actual number, how does it help you? It isn't even practical to sum up a couple of dozen terms for anyone who wants to go about the ordinary business of living; how, then, can it be possible to sum up an infinite number?

Ah, but mathematicians do not give up on the sum of a series just because the number of terms in it is unending. For instance, the series:

$$\tfrac{1}{2} + \tfrac{1}{4} + \tfrac{1}{8} + \tfrac{1}{16} + \tfrac{1}{32} + \tfrac{1}{64} \cdots$$

can be summed up, using successively more and more terms. If you do this, you will find that the more terms you use, the closer you get to 1, and you can express this in shorthand form by saying that the sum of that infinite number of terms is merely 1 after all.

There is a formula, in fact, that can be used to determine the sum of any decreasing geometric progression, of which the above is an example.

Thus, the series:

$$\frac{3}{10} + \frac{3}{100} + \frac{3}{1000} + \frac{3}{10000} + \frac{3}{100000} \cdots$$

adds up, in all its splendidly infinite numbers, to a mere $\frac{1}{3}$, and the series:

$$\frac{1}{2} + \frac{1}{20} + \frac{1}{200} + \frac{1}{2000} + \frac{1}{20000} \cdots$$

adds up to $\frac{5}{9}$.

To be sure, the series worked out for the evaluation of π are none of them decreasing geometric progressions, and so the formula cannot be used to evaluate the sum. In fact, no formula has ever been found to evaluate the sum of the Leibniz series or any of the others. Nevertheless, there seemed no reason at first to suppose that there might not be some way of finding a decreasing geometric progression that would evaluate π. If so, π would then be expressible as a fraction. A fraction is actually the ratio of two numbers and anything expressible as a fraction, or ratio, is a "rational number," as I explained in the previous chapter. The hope, then, was that π might be a rational number.

One way of proving that a quantity is a rational number is to work out its value decimally as far as you can (by adding up more and more terms of an infinite series, for instance) and then show the result to be a "repeating decimal"; that is, a decimal in which digits or some group of digits repeat themselves endlessly.

For instance, the decimal value of $\frac{1}{3}$ is 0.33333333333 . . . , while that of $\frac{1}{7}$ is 0.142857 142857 142857 . . . , and so on endlessly. Even a fraction such as $\frac{1}{8}$ which seems to "come out even" is really a repeating decimal if you count zeros, since its decimal equivalent is 0.125000-000000. . . . It can be proved mathematically that every fraction, however complicated, can be expressed as a decimal which sooner or later becomes a repeating one. Conversely, any decimal which ends by becoming a repeating one, however involved the repetitive cycle, can be expressed as an exact fraction.

Take any repeating decimal at random, say 0.373737-37373737. . . . First, you can make a decreasing geometrical progression out of it by writing it as:

$$\frac{37}{100} + \frac{37}{10000} + \frac{37}{1000000} + \frac{37}{100000000} \cdots$$

and you can then use the formula to work out its sum, which comes out to $\frac{37}{99}$. (Work out the decimal equivalent of that fraction and see what you get.)

Or suppose you have a decimal which starts out non-repetitively and then becomes repetitive, such as 15.2165-5555555555. . . . This can be written as:

$$15 + \frac{216}{1000} + \frac{5}{10000} + \frac{5}{100000} + \frac{5}{1000000} \cdots$$

From $\frac{5}{10000}$ on, we have a decreasing geometric progression and its sum works out to be $\frac{5}{9000}$. So the series becomes a finite one made out of exactly three terms and no more, and can be summed easily:

$$15 + \frac{216}{1000} + \frac{5}{90000} = \frac{136949}{9000}$$

If you wish, work out the decimal equivalent of $\frac{136949}{9000}$ and see what you get.

Well, then, if the decimal equivalent of π were worked out for a number of decimal places and some repetition were discovered in it, however slight and however complicated, provided it could be shown to go on endlessly, a new series could be written to express its exact value. This new series would conclude with a decreasing geometric progression which could be summed. There would then be a finite series and the true value of π could be expressed not as a series but as an actual number.

Mathematicians threw themselves into the pursuit. In 1593 Vieta himself used his own series to calculate π to seventeen decimal places. Here it is, if you want to stare at it: 3.14159265358979323. As you see, there are no apparent repetitions of any kind.

Then in 1615 the German mathematician Ludolf von Ceulen used an infinite series to calculate π to thirty-five places. He found no signs of repetitiveness, either. However, this was so impressive a feat for his time that he won a kind of fame, for π is sometimes called "Ludolf's number" in consequence, at least in German textbooks.

And then in 1717 the English mathematician Abraham Sharp went Ludolf several better by finding π to seventy-two decimal places. Still no sign of repeating.

But shortly thereafter, the game was spoiled.

To prove a quantity is rational, you have to present the fraction to which it is equivalent and display it. To prove it is irrational, however, you need not necessarily work out a single decimal place. What you must do is to *suppose* that the quantity can be expressed by a fraction, $\frac{p}{q}$, and then demonstrate that this involves a contradiction, such as that p must at the same time be even and odd. This would prove that no fraction could express the quantity, which would therefore be irrational.

Exactly this sort of proof was developed by the ancient Greeks to show that the square root of 2 was an irrational number (the first irrational ever discovered). The Pythagoreans were supposed to have been the first to discover this and to have been so appalled at finding that there could be quantities that could not be expressed by any fraction, however complicated, that they swore themselves to secrecy and provided a death penalty for snitching. But like all scientific secrets, from irrationals to atom bombs, the information leaked out anyway.

Well, in 1761 a German physicist and mathematician Johann Heinrich Lambert finally proved that π was irrational. Therefore, no pattern at all was to be expected, no matter how slight and no matter how many decimal places were worked out. The true value can *only* be expressed as an infinite series.

Alas!

But shed no tears. Once π was proved irrational, mathematicians were satisfied. The problem was over. And as for the application of π to physical calculations, that problem was over and done with, too. You may think that sometimes in very delicate calculations it might be necessary to know π to a few dozen or even to a few hundred places, but not so! The delicacy of scientific measurements is wonderful these days, but still there are few that approach, say, one part in a billion, and for anything that accurate which involves the use of π, nine or ten decimal places would be ample.

For example, suppose you drew a circle ten billion miles across, with the sun at the center, for the purpose of enclosing the entire solar system, and suppose you wanted to calculate the length of the circumference of this circle (which would come to over thirty-one billion miles) by using $\frac{355}{113}$ as the approximate value of π. You would be off by less than three thousand miles.

But suppose you were so precise an individual that you found an error of three thousand miles in 31,000,000,000 to be insupportable. You might then use Ludolf's value of π to thirty-five places. You would then be off by a distance that would be equivalent to a millionth of the diameter of a proton.

Or let's take a *big* circle, say the circumference of the known universe. Large radio telescopes under construction will, it is hoped, receive signals from a distance as great as 40,000,000,000 light-years. A circle about a universe with such a radius would have a length of, roughly, 150,000,-000,000,000,000,000,000 (150 sextillion) miles. If the length of this circumference were calculated by Ludolf's value of π to thirty-five places, it would be off by less than a millionth of an inch.

What can one say then about Sharp's value of π to seventy-two places?

Obviously, the value of π, as known by the time its irrationality was proven, was already far beyond the accuracy that could conceivably be demanded by science, now or in the future.

And yet with the value of π no longer needed for scientists, past what had already been determined, people nevertheless continued their calculations through the first half of the nineteenth century.

A fellow called George Vega got π to 140 places, another called Zacharias Dase did it to 200 places, and someone called Recher did it to 500 places.

Finally, in 1873, William Shanks reported the value of π to 707 places, and that, until 1949, was the record—and small wonder. It took Shanks fifteen years to make the calculation and, for what that's worth, no signs of any repetitiveness showed up.

We can wonder about the motivation that would cause a man to spend fifteen years on a task that can serve no purpose. Perhaps it is the same mental attitude that will make a man sit on a flagpole or swallow goldfish in order to "break a record." Or perhaps Shanks saw this as his one road to fame.

If so, he made it. Histories of mathematics, in among their descriptions of the work of men like Archimedes, Fermat, Newton, Euler, and Gauss, will also find room for a line to the effect that William Shanks in the years preceding 1873 calculated π to 707 decimal places. So perhaps he felt that his life had not been wasted.

But alas, for human vanity—

In 1949 the giant computers were coming into their own, and occasionally the young fellows at the controls, full of fun and life and beer, could find time to play with them.

So, on one occasion, they pumped one of the unending series into the machine called ENIAC and had it calculate the value of π. They kept it at the task for seventy hours, and at the end of that time they had the value of π (shades of Shanks!) to 2035 places.[1]

And to top it all off for poor Shanks and his fifteen wasted years, an error was found in the five hundred umpty-umpth digit of Shanks' value, so that all the digits after that, well over a hundred, were *wrong!*

And of course, in case you're wondering, and you shouldn't, the values as determined by computers showed no signs of any repetitiveness either.

5. Tools of the Trade

THE PREVIOUS CHAPTER does not conclude the story of π. As the title stated, it was only a piece of π. Let us therefore continue onward.

The Greek contribution to geometry consisted of idealizing and abstracting it. The Egyptians and Babylonians

---

[1] By 1955 a faster computer calculated π to 10,017 places in thirty-three hours and, actually, there *are* interesting mathematical points to be derived from studying the various digits of π.

solved specific problems by specific methods but never tried to establish general rules.

The Greeks, however, strove for the general and felt that mathematical figures had certain innate properties that were eternal and immutable. They felt also that a consideration of the nature and relationships of these properties was the closest man could come to experiencing the sheer essence of beauty and divinity. (If I may veer away from science for a moment and invade the sacred precincts of the humanities, I might point out that just this notion was expressed by Edna St. Vincent Millay in a famous line that goes: "Euclid alone has looked on Beauty bare.")

Well, in order to get down to the ultimate bareness of Beauty, one had to conceive of perfect, idealized figures made up of perfect idealized parts. For instance, the ideal line consisted of length and nothing else. It had neither thickness nor breadth nor anything, in fact, but length. Two ideal lines, ideally and perfectly straight, intersected at an ideal and perfect point, which had no dimensions at all, only position. A circle was a line that curved in perfectly equal fashion at all points; and every point on that curve was precisely equally distant from a particular point called the center of the circle.

Unfortunately, although one can imagine such abstractions, one cannot communicate them as abstractions alone. In order to explain the properties of such figures (and even in order to investigate them on your own) it is helpful, almost essential in fact, to draw crass, crude, and ungainly approximations in wax, on mud, on blackboard, or on paper, using a pointed stick, chalk, pencil, or pen. (Beauty must be swathed in drapery in mathematics, alas, as in life.)

Furthermore, in order to prove some of the ineffably beautiful properties of various geometrical figures, it was usually necessary to make use of more lines than existed in the figure alone. It might be necessary to draw a new line through a point and make it parallel or, perhaps, perpendicular to a second line. It might be necessary to divide a line into equal parts, or to double the size of an angle.

To make all this drawing as neat and as accurate as possible, instruments must be used. It follows naturally, I think, once you get into the Greek way of thinking, that the fewer and simpler the instruments used for the purpose, the closer the approach to the ideal.

Eventually, the tools were reduced to an elegant minimum of two. One is a straightedge for the drawing of straight lines. This is not a ruler, mind you, with inches or centimeters marked off on it. It is an unmarked piece of wood (or metal or plastic, for that matter) which can do no more than guide the marking instrument into the form of a straight line.

The second tool is the compass, which, while most simply used to draw circles, will also serve to mark off equal segments of lines, will draw intersecting arcs that mark a point that is equidistant from two other points, and so on.

I presume most of you have taken plane geometry and have utilized these tools to construct one line perpendicular to another, to bisect an angle, to circumscribe a circle about a triangle, and so on. All these tasks and an infinite number of others can be performed by using the straightedge and compass in a finite series of manipulations.

By Plato's time, of course, it was known that by using more complex tools, certain constructions could be simplified; and, in fact, that some constructions could be performed which, until then, could not be performed by straightedge and compass alone. That, to the Greek geometers, was something like shooting a fox or a sitting duck, or catching fish with worms, or looking at the answers in the back of the book. It got results but it just wasn't the gentlemanly thing to do. The straightedge and compass were the only "proper" tools of the geometrical trade.

Nor was it felt that this restriction to the compass and straightedge unduly limited the geometer. It might be tedious at times to stick to the tools of the trade; it might be easier to take a short cut by using other devices; but surely the straightedge and compass alone could do it all, if you were only persistent enough and ingenious enough.

For instance, if you are given a line of a fixed length which is allowed to represent the numeral 1, it is possible to construct another line, by compass and straightedge alone, exactly twice that length to represent 2, or another line to represent 3 or 5 or 500 or $\frac{1}{2}$ or $\frac{1}{3}$ or $\frac{1}{5}$ or $\frac{3}{5}$ or

$2\frac{3}{5}$ or $27\frac{16}{23}$. In fact, by using compass and straightedge only, any rational number (i.e., any integer or fraction) could be duplicated geometrically. You could even make use of a simple convention (which the Greeks never did, alas) to make it possible to represent both positive and negative rational numbers.

Once irrational numbers were discovered, numbers for which no definite fraction could be written, it might seem that compass and straightedge would fail, but even then they did not.

For instance, the square root of 2 has the value 1,414214 . . . and on and on without end. How, then, can you construct one line which is 1.414214 . . . times as long as another when you cannot possibly ever know exactly how many times as long you want it to be.

Actually, it's easy. Imagine a given line from point A to point B. (I can do this without a diagram, I think, but if you feel the need you can sketch the lines as you read. It won't be hard.) Let this line, AB, represent L.

Next, construct a line at B, perpendicular to AB. Now you have two lines forming a right angle. Use the compass to draw a circle with its center at B, where the two lines meet, and passing through A. It will cut the perpendicular line you have just drawn at a point we can call C. Because of the well-known properties of the circle, line BC is exactly equal to line AB, and is also 1.

Finally, connect points A and C with a third straight line.

That line, AC, as can be proven by geometry, is exactly $\sqrt{2}$ times as long as either AB or BC, and therefore represents the irrational quantity $\sqrt{2}$.

Don't, of course, think that it is now only necessary to measure AC in terms of AB to obtain an exact value of $\sqrt{2}$. The construction was drawn by imperfect instruments in the hands of imperfect men and is only a crude approximation of the ideal figures they represent. It is the ideal line represented by AC that is $\sqrt{2}$, and not AC itself in actual reality.

It is possible, in similar fashion, to use the straightedge and compass to represent an infinite number of other irrational quantities.

In fact, the Greeks had no reason to doubt that any

conceivable number at all could be represented by a line that could be constructed by use of straightedge and compass alone in a finite number of steps. And since all constructions boiled down to the construction of certain lines representing certain numbers, it was felt that anything that could be done with any tool could be done by straightedge and compass alone. Sometimes the details of the straightedge and compass construction might be elusive and remain undiscovered, but eventually, the Greeks felt, given enough ingenuity, insight, intelligence, intuition, and luck, the construction could be worked out.

For instance, the Greeks never learned how to divide a circle into seventeen equal parts by straightedge and compass alone. Yet it could be done. The method was not discovered until 1801, but in that year, the German mathematician Karl Friedrich Gauss, then only twenty-four, managed it. Once he divided the circle into seventeen parts, he could connect the points of division by a straightedge to form a regular polygon of seventeen sides (a "septendecagon"). The same system could be used to construct a regular polygon of 257 sides, and an infinite number of other polygons with still more sides, the number of sides possible being calculated by a formula which I won't give here.

If the construction of a simple thing like a regular septendecagon could elude the great Greek geometers and yet be a perfectly soluble problem in the end, why could not any conceivable construction, however puzzling it might seem, yet prove soluble in the end.

As an example, one construction that fascinated the Greeks was this: Given a circle, construct a square of the same area.

This is called "squaring the circle."

There are several ways of doing this. Here's one method. Measure the radius of the circle with the most accurate measuring device you have—say, just for fun, that the radius proves to be one inch long precisely. (This method will work for a radius of any length, so why not luxuriate in simplicity.) Square that radius, leaving the value still 1, since 1×1 is 1, thank goodess, and multiply that by the best value of π you can find. (Were you wondering when I'd get back to π?) If you use 3,1415926 as your value of

π, the area of the circle proves to be 3.14115926 square inches.

Now, take the square root of that, which is 1.7724539 inches, and draw a straight line exactly 1.7724539 inches long, using your measuring device to make sure of the length. Construct a perpendicular at each end of the line, mark off 1.7724539 inches on each perpendicular, and connect those two points.

Voilà! You have a square equal in area to the given circle. Of course, you may feel uneasy. Your measuring device isn't infinitely accurate and neither is the value of π which you used. Does not this mean that the squaring of the circle is only approximate and not exact?

Yes, but it is not the details that count but the principle. We can *assume* the measuring device to be perfect, and the value of π which was used to be accurate to an infinite number of places. After all, this is just as justifiable as assuming our actual drawn lines to represent ideal lines, considering our straightedge perfectly straight and our compass to end in two perfect points. In principle, we have indeed perfectly squared the circle.

Ah, but we have made use of a measuring device, which is not one of the only two tools of the trade allowed a gentleman geometer. That marks you as a cad and bounder and you are hereby voted out of the club.

Here's another method of squaring the circle. What you really need, assuming the radius of your circle to represent 1, is another straight line representing $\sqrt{\pi}$. A square built on such a line would have just the area of a unit-radius circle. How to get such a line? Well, if you could construct a line equal to π times the length of the radius, there are known methods, using straightedge and compass alone, to construct a line equal in length to the square root of that line, hence representing the $\sqrt{\pi}$, which we are after.

But it is simple to get a line that is π times the radius. According to a well-known formula, the circumference of the circle is equal in length to twice the radius times π. So let us imagine the circle resting on a straight line and let's make a little mark at the point where the circle just touches the line. Now slowly turn the circle so that it moves along the line (without slipping) until the point you have marked makes a complete circuit and once again touches the line. Make another mark where it again touches. Thus, you have marked off the circumference of the circle on a straight line and the distance between the two marks is twice π.

Bisect that marked-off line by the usual methods of straightedge and compass geometry and you have a line representing π. Construct the square root of that line and you have $\sqrt{\pi}$.

Voilà! By that act, you have, in effect, squared the circle.

But no. I'm afraid you're still out of the club. You have made use of a rolling circle with a mark on it and that comes under the heading of an instrument other than the straightedge and compass.

The point is that there are any number of ways of squaring the circle, but the Greeks were unable to find any way of doing it with straightedge and compass alone in a finite number of steps. (They spent I don't know how many man-hours of time searching for a method, and looking back on it, it might all seem an exercise in futility now, but it wasn't. In their search, they came across all sorts of new curves, such as the conic sections, and new theorems, which were far more valuable than the squaring of the circle would have been.)

Although the Greeks failed to find a method, the search continued and continued. People kept on trying and trying and trying and trying—

And now let's change the subject for a while.

Consider a simple equation such as $2x - 1 = 0$. You can see that setting $x = \frac{1}{2}$ will make a true statement out of it, for $2(\frac{1}{2}) - 1$ is indeed equal to zero. No other number can be substituted for x in this equation and yield a true statement.

By changing the integers in the equation (the "coefficients" as they are called) x can be made to equal other specific numbers. For instance, in $3x - 4 = 0$, x is equal to $\frac{4}{3}$; and in $7x + 2 = 0$, $x - -\frac{2}{7}$. In fact, by choosing the coefficients appropriately, you can have as a value of x any positive or negative integer or fraction whatever.

But in such an "equation of the first degree," you can only obtain rational values for x. You can't possibly have an equation of the form $Ax + B = 0$, where A and B are

64

rational, such that x will turn out to be equal to $\sqrt{2}$, for instance.

The thing to do is to try a more complicated variety of equation. Suppose you try $x^2 - 2 = 0$, which is an "equation of the second degree" because it involves a square. If you solve for x you'll find the answer, $\sqrt{2}$, when substituted for x will yield a true statement. In fact, there are two possible answers, for the substitution of $-\sqrt{2}$ for x will also yield a true statement.

You can build up equations of the third degree, such as $Ax^3 + Bx^2 + Cx + D = 0$, or of the fourth degree (I don't have to give any more examples, do I?), or higher. Solving for x in each case becomes more and more difficult, but will give solutions involving cube roots, fourth roots, and so on.

In any equation of this type (a "polynomial equation") the value of x can be worked out by manipulating the coefficients. To take the simplest case, in the general equation of the first degree: $Ax + B = 0$, the value of x is $-B/A$. In the general equation of the second degree: $Ax^2 + Bx + C = 0$, there are two solutions. One is $\dfrac{-B + \sqrt{B^2 - 4AC}}{2A}$ and the other is $\dfrac{-B - \sqrt{B^2 - 4AC}}{2A}$.

Solutions get progressively more complicated and eventually, for equations of the fifth degree and higher, no general solution can be given, although specific solutions can still be worked out. The principle remains, however, that in all polynomial equations, the value of x can be expressed by use of a finite number of integers involved in a finite number of operations, these operations consisting of addition, subtraction, multiplication, division, raising to a power ("involution"), and extracting roots ("evolution").

These operations are the only ones used in ordinary algebra and are therefore called "algebraic operations." Any number which can be derived from the integers by a finite number of algebraic operations in any combination is called an "algebraic number." To put it in reverse, any algebraic number is a possible solution for some polynomial equation.

Now it so happens that the geometric equivalent of all the algebraic operations, except the extraction of roots

65

higher than the square root, can be performed by straightedge and compass alone. If a given line represents 1, therefore, it follows that a line representing *any* algebraic number that involves no root higher than the square root can be constructed by straightedge and compass in a finite number of manipulations.

Since π does not seem to contain any cube roots (or worse), is it possible that it can be constructed by straightedge and compass? That might be if algebraic numbers included *all* numbers. But do they? Are there numbers which cannot be solutions to any polynomial equation, and are therefore not algebraic?

To begin with, all possible rational numbers can be solutions to equations of the first degree, so all rational numbers are algebraic numbers. Then, certainly some irrational numbers are algebraic numbers, for it is easy to write equations for which $\sqrt{2}$ or $\sqrt[5]{15} - 3$ are solutions.

But can there be irrational numbers which will not serve as a solution to a single one of the infinite numbers of different polynomial equations in each of all the infinite number of degrees possible?

In 1844 the French mathematician Joseph Liouville finally found a way of showing that such nonalgebraic numbers did exist. (No, I don't know how he did it, but if any reader thinks I can understand the method, and I must warn him not to overestimate me, he is welcome to send it in.)

However, having proved that nonalgebraic numbers existed, Liouville could still not find a specific example. The nearest he came was to show that a number represented by the symbol *e* could not serve as the root for any conceivable equation of the second degree.

(At this point I am tempted to launch into a discussion of the number *e* because, as I said at the start of the previous chapter, there is the famous equation $e^{\pi i} = -1$. But I'll resist temptation. I'll say only that *e* is an irrational number with a value that has now been calculated to sixty thousand places, of which the first twenty-five decimals are: 2.7182818284590452353602874.)

Then, in 1873, the French mathematician Charles Hermite worked out a method of analysis that showed that *e* could not be the root of any conceivable equation of any conceivable degree and hence was actually not an algebraic number. It was, in fact, what is called a "transcendental number," one which transcends (that is, goes beyond) the algebraic operations and cannot be produced from the integers by any finite number of those operations. (That is, $\sqrt{2}$ is irrational but can be produced by a single algebraic operation, taking the square root of 2. The value of e, on the other hand, can only be calculated by the use of infinite series involving an infinite number of additions, divisions, subtractions, and so on.)

Using the methods developed by Hermite, the German mathematician Ferdinand Lindemann in 1882 proved that π, too, was a transcendental number.

This is crucial for the purposes of this article, for it meant that a line segment equivalent to π cannot be built up by the use of the straightedge and compass alone in a finite number of manipulations. *The circle cannot be squared by straightedge and compass alone.* It is as impossible to do this as to find an exact value for $\sqrt{2}$, or to find an odd number that is an exact multiple of 4.

One odd point about transcendental numbers—
They were difficult to find, but now that they have been, they prove to be present in overwhelming numbers. Practically any expression that involves either e or π is transcendental, provided the expression is not arranged so that the e or π cancel out. Practically all expressions involving logarithms (which involve e) and practically all expressions involving trigonometric functions (which involve π) are transcendental. Expressions involving numbers raised to an irrational power, such as $x\sqrt{2}$, are transcendental.

In fact, if you refer back to Chapter 3, you will understand me when I say that it has been proved that the algebraic numbers can be put into one-to-one correspondence with the integers, but the transcendental numbers can not.

This means that the algebraic numbers, although infinite, belong to the lowest of the transfinite numbers, $\aleph_0$, while the transcendental numbers belong to the next higher transfinite, $\aleph_1$. There are thus infinitely more transcendental numbers than there are algebraic numbers.

To be sure, the fact that the transcendentality of π is now well established and has been for nearly a century doesn't stop the ardent circle-squarers, who continue to work away desperately with straightedge and compass and continue to report solutions regularly.

So if *you* know a way to square the circle by straightedge and compass alone, I congratulate you, but you have a fallacy in your proof somewhere. And it's no use sending it to me, because I'm a rotten mathematician and couldn't possibly find the fallacy, but I tell you anyway, it's there.

6. The Imaginary That Isn't

WHEN I was a mere slip of a lad and attended college, I had a friend with whom I ate lunch every day. His 11 A.M. class was in sociology, which I absolutely refused to take, and my 11 A.M. class was calculus, which he as steadfastly refused to take—so we had to separate at eleven and meet at twelve.

As it happened, his sociology professor was a scholar who did things in the grand manner, holding court after class was over. The more eager students gathered close and listened to him pontificate for an additional fifteen minutes, while they threw in an occasional log in the form of a question to feed the flame of oracle.

Consequently, when my calculus lecture was over, I had to enter the sociology room and wait patiently for court to conclude.

Once I walked in when the professor was listing on the board his classification of mankind into the two groups of mystics and realists, and under mystics he had included the mathematicians along with the poets and theologians. One student wanted to know why.

"Mathematicians," said the professor, "are mystics because they believe in numbers that have no reality."

Now ordinarily, as a nonmember of the class, I sat in

the corner and suffered in silent boredom, but now I rose convulsively, and said, "What numbers?"

The professor looked in my direction and said, "The square root of minus one. It has no existence. Mathematicians call it imaginary. But they believe it has some kind of existence in a mystical way."

"There's nothing mystical about it," I said, angrily. "The square root of minus one is just as real as any other number."

The professor smiled, feeling he had a live one on whom he could now proceed to display his superiority of intellect (I have since had classes of my own and I know exactly how he felt). He said, silkily, "We have a young mathematician here who wants to prove the reality of the square root of minus one. Come, young man, hand me the square root of minus one pieces of chalk!"

I reddened, "Well, now, wait—"

"That's all," he said, waving his hand. Mission, he imagined, accomplished, both neatly and sweetly.

But I raised my voice. "I'll do it. I'll do it. I'll hand you the square root of minus one pieces of chalk, if you hand me a one-half piece of chalk."

The professor smiled again, and said, "Very well," broke a fresh piece of chalk in half, and handed me one of the halves. "Now for your end of the bargain."

"Ah, but wait," I said, "you haven't fulfilled your end. This is one piece of chalk you've handed me, not a one-half piece." I held it up for the others to see. "Wouldn't you all say this was one piece of chalk? It certainly isn't two or three."

Now the professor wasn't smiling. "Hold it. One piece of chalk is a piece of regulation length. You have one that's half the regulation length."

I said, "Now you're springing an arbitrary definition on me. But even if I accept it, are you willing to maintain that this is a one-half piece of chalk and not a 0.48 piece or a 0.52 piece? And can you really consider yourself qualified to discuss the square root of minus one, when you're a little hazy on the meaning of one half?"

But by now the professor had lost his equanimity altogether and his final argument was unanswerable. He said, "Get the hell out of here!" I left (laughing) and thereafter waited for my friend in the corridor.

Twenty years have passed since then and I suppose I ought to finish the argument—

Let's start with a simple algebraic equation such as $x + 3 = 5$. The expression, x, represents some number which, when substituted for x, makes the expression a true equality. In this particular case x must equal 2, since $2 + 3 = 5$, and so we have "solved for x."

The interesting thing about this solution is that it is the *only* solution. There is no number but 2 which will give 5 when 3 is added to it.

This is true of any equation of this sort, which is called a "linear equation" (because in geometry it can be represented as a straight line) or "a polynomial equation of the first degree." No polynomial equation of the first degree can ever have more than one solution for x.

There are other equations, however, which *can* have more than one solution. Here's an example: $x^2 - 5x + 6 = 0$, where x^2 ("x square" or "x squared") represents x times x. This is called a "quadratic equation," from a Latin word for "square," because it involves x square. It is also called "a polynomial equation of the second degree" because of the little ^{2} in x^2. As for x itself, that could be written x^1, except that the ^{1} is always omitted and taken for granted, and that is why $x + 3 = 5$ is an equation of the first degree.

If we take the equation $x^2 - 5x + 6 = 0$, and substitute 2 for x, then x^2 is 4, while $5x$ is 10, so that the equation becomes $4 - 10 + 6 = 0$, which is correct, making 2 a solution of the equation.

However, if we substitute 3 for x, then x^2 is 9 and $5x$ is 15, so that the equation becomes $9 - 15 + 6 = 0$, which is also correct, making 3 a second solution of the equation.

Now no equation of the second degree has ever been found which has more than two solutions, but what about polynomial equations of the third degree? These are equations containing x^3 ("x cube" or "x cubed"), which are therefore also called "cubic equations." The expression x^3 represents x times x times x.

The equation $x^3 - 6x^2 + 11x - 6 = 0$ has three solutions, since you can substitute 1, 2, or 3 for x in this equation and come up with a true equality in each case. No cubic equation has ever been found with more than three solutions, however.

In the same way polynomial equations of the fourth degree can be constructed which have four solutions but no more; polynomial equations of the fifth degree, which have five solutions but no more; and so on. You might say, then, that a polynomial equation of the nth degree can have as many as n solutions, but never more than n.

Mathematicians craved something even prettier than that and by about 1800 found it. At that time, the German mathematician Karl Friedrich Gauss showed that every equation of the nth degree had exactly n solutions, not only no more, but also no less.

However, in order to make the fundamental theorem true, our notion of what constitutes a solution to an algebraic equation must be drastically enlarged.

To begin with, men accept the "natural numbers" only: 1, 2, 3, and so on. This is adequate for counting objects that are considered as units generally. You can have 2 children, 5 cows, or 8 pots; while to have 2½ children, 5¼ cows, or 8⅓ pots does not make much sense.

In measuring continuous quantities such as lengths or weights, however, fractions became essential. The Egyptians and Babylonians managed to work out methods of handling fractions, though these were not very efficient by our own standards; and no doubt conservative scholars among them sneered at the mystical mathematicians who believed in a number like 5½, which was neither 5 nor 6.

Such fractions are really ratios of whole numbers. To say a plank of wood is 2⅝ yards long, for instance, is to say that the length of the plank is to the length of a standard yardstick as 21 is to 8. The Greeks, however, discovered that there were definite quantities which could not be expressed as ratios of whole numbers. The first to be discovered was the square root of 2, commonly expressed as $\sqrt{2}$, which is that number which, when multiplied by itself, gives 2. There is such a number but it cannot be expressed as a ratio; hence, it is an "irrational number."

Only thus far did the notion of number extend before modern times. Thus, the Greeks accepted no number smaller than zero. How can there be less than nothing?

To them, consequently, the equation $x + 5 = 3$ had no solution. How can you add 5 to any number and have 3 as a result? Even if you added 5 to the smallest number (that is, to zero), you would have 5 as the sum, and if you added 5 to any other number (which would have to be larger than zero), you would have a sum greater than 5.

The first mathematician to break this taboo and make systematic use of numbers less than zero was the Italian, Girolamo Cardano. After all, there *can* be less than nothing. A debt is less than nothing.

If all you own in the world is a two-dollar debt, you have two dollars less than nothing. If you are then given five dollars, you end with three dollars of your own (assuming you are an honorable man who pays his debts). Consequently, in the equation $x + 5 = 3$, x can be set equal to -2, where the minus sign indicates a number less than zero.

Such numbers are called "negative numbers," from a Latin word meaning "to deny," so that the very name carries the traces of the Greek denial of the existence of such numbers. Numbers greater than zero are "positive numbers" and these can be written $+1$, $+2$, $+3$, and so on.

From a practical standpoint, extending the number system by including negative numbers simplifies all sorts of computations; as, for example, those in bookkeeping.

From a theoretical standpoint, the use of negative numbers means that every equation of the first degree has exactly one solution. No more; no less.

If we pass on to equations of the second degree, we find that the Greeks would agree with us that the equation $x^2 - 5x + 6 = 0$ has two solutions, 2 and 3. They would say, however, that the equation $x^2 + 4x - 5 = 0$ has only one solution, 1. Substitute 1 for x and x^2 is 1, while $4x$ is 4, so that the equation becomes $1 + 4 - 5 = 0$. No other number will serve as a solution, as long as you restrict yourself to positive numbers.

However, the number -5 is a solution, if we consider a few rules that are worked out in connection with the multiplication of negative numbers. In order to achieve consistent results, mathematicians have decided that the multiplication of a negative number by a positive number yields a negative product, while the multiplication of a negative number by a negative number yields a positive product.

If, in the equation $x^2 + 4x - 5 = 0$, -5 is substituted for x, then x^2 becomes -5 times -5, or $+25$, while $4x$ becomes $+4$ times -5, or -20. The equation becomes $25 - 20 - 5 = 0$, which is true. We would say, then, that there are two solutions to this equation, $+1$ and -5.

Sometimes, a quadratic equation does indeed seem to have but a single root, as, for example, $x^2 - 6x + 9 = 0$, which will be a true equality if and only if the number $+3$ is substituted for x. However, the mechanics of solution of the equation show that there are actually two solutions, which happen to be identical. Thus, $x^2 - 6x + 9 = 0$ can be converted to $(x - 3)(x - 3) = 0$ and each $(x - 3)$ yields a solution. The two solutions of this equation are, therefore, $+3$ and $+3$.

Allowing for occasional duplicate solutions, are we ready to say then that all second-degree equations can be shown to have exactly two solutions if negative numbers are included in the number system?

Alas, no! For what about the equation $x^2 + 1 = 0$. To begin with, x^2 must be -1, since substituting -1 for x^2 makes the equation $-1 + 1 = 0$, which is correct enough.

But if x^2 is -1, then x must be the famous square root of minus one ($\sqrt{-1}$), which occasioned the set-to between the sociology professor and myself. The square root of minus one is that number which when multiplied by itself will give -1. But there is no such number in the set of positive and negative quantities, and that is the reason the sociology professor scorned it. First, $+1$ times $+1$ is $+1$; secondly, -1 times -1 is $+1$.

To allow any solution at all for the equation $x^2 + 1 = 0$, let alone two solutions, it is necessary to get past this roadblock. If no positive number will do and no negative one either, it is absolutely essential to define a completely new kind of number; an imaginary number, if you like; one with its square equal to -1.

We could, if we wished, give the new kind of number a special sign. The plus sign does for positives and the minus sign for negatives; so we could use an asterisk for the new number and say that *1 ("star one") times *1 was equal to -1.

However, this was not done. Instead, the symbol i (for "imaginary") was introduced by the Swiss mathematician Leonhard Euler in 1777 and was thereafter generally adopted. So we can write $i = \sqrt{-1}$ or $i^2 = -1$.

Having defined i in this fashion, we can express the square root of any negative number. For instance, $\sqrt{-4}$ can be written $\sqrt{4}$ times $\sqrt{-1}$, or $2i$. In general, any square root of a negative number, $\sqrt{-n}$, can be written as the square root of the equivalent positive number times the square root of minus one; that is, $\sqrt{-n} = \sqrt{n}i$.

In this way, we can picture a whole series of imaginary numbers exactly analogous to the series of ordinary or "real numbers." For 1, 2, 3, 4, . . . , we would have i, $2i$, $3i$, $4i$. . . . This would include fractions, for ⅔ would be matched by $\frac{2i}{3}$; $\frac{15}{17}$ by $\frac{15i}{17}$, and so on. It would also include irrationals, for $\sqrt{2}$ would be matched by $\sqrt{2}i$ and even a number like π (pi) would be matched by πi.

These are all comparisons of positive numbers with imaginary numbers. What about negative numbers? Well, why not negative imaginaries, too? For -1, -2, -3, -4, . . . , there would be $-i$, $-2i$, $-3i$, $-4i$. . . .

So now we have four classes of numbers: 1) positive real numbers, 2) negative real numbers, 3) positive imaginary numbers, 4) negative imaginary numbers. (When a negative imaginary is multiplied by a negative imaginary, the product is negative.)

Using this further extension of the number system, we can find the necessary two solutions for the equation $x^2 + 1 = 0$. They are $+i$ and $-i$. First $+i$ times $+i$ equals -1, and secondly $-i$ times $-i$ equals -1, so that in either case, the equation becomes $-1 + 1 = 0$, which is a true equality.

In fact, you can use the same extension of the number system to find all four solutions for an equation such as $x^4 - 1 = 0$. The solutions are $+1$, -1, $+i$, and $-i$. To show this, we must remember that any number raised to the fourth power is equal to the square of that number multiplied by itself. That is, n^4 equals n^2 times n^2.

Now let's substitute each of the suggested solutions into the equations so that x^4 becomes successively $(+1)^4$, $(-1)^4$, $(+i)^4$, and $(-i)^4$.

First $(+1)^4$ equals $(+1)^2$ times $(+1)^2$, and since $(+1)^2$ equals $+1$, that becomes $+1$ times $+1$, which is $+1$.

Second, $(-1)^4$ equals $(-1)^2$ times $(-1)^2$, and since $(-1)^2$ also equals $+1$, the expression is again $+1$ times $+1$, or $+1$.

Third, $(+i)^4$ equals $(+i)^2$ times $(+i)^2$ and we have defined $(+i)^2$ as -1, so that the expression becomes -1 times -1, or $+1$.

Fourth, $(-i)^4$ equals $(-i)^2$ times $(-i)^2$, which is also -1 times -1, or $+1$.

All four suggested solutions, when substituted into the equation $x^4 - 1 = 0$, give the expression $+1 - 1 = 0$, which is correct.

It might seem all very well to talk about imaginary numbers—for a mathematician. As long as some defined quantity can be made subject to rules of manipulation that do not contradict anything else in the mathematical system, the mathematician is happy. He doesn't really care what it "means."

Ordinary people do, though, and that's where my sociologist's charge of mysticism against mathematicians arises.

And yet it is the easiest thing in the world to supply the so-called "imaginary" numbers with a perfectly real and concrete significance. Just imagine a horizontal line crossed by a vertical line and call the point of intersection zero. Now you have four lines radiating out at mutual right angles from that zero point. You can equate those lines with the four kinds of numbers.

If the line radiating out to the right is marked off at equal intervals, the marks can be numbered $+1$, $+2$, $+3$, $+4$, . . . , and so on for as long as we wish, if we only make the line long enough. Between the markings are all the fractions and irrational numbers. In fact, it can be shown that to every point on such a line there corresponds one and only one positive real number, and for every positive real number there is one and only one point on the line.

The line radiating out to the left can be similarly marked off with the negative real numbers, so that the horizontal line can be considered the "real-number axis," including both positives and negatives.

Similarly, the line radiating upward can be marked off

with the positive imaginary numbers, and the one radiating downward with the negative imaginary numbers. The vertical line is then the imaginary-number axis.

Suppose we label the different numbers not by the usual signs and symbols, but by the directions in which the lines point. The rightward line of positive real numbers can be called East because that would be its direction of extension on a conventional map. The leftward line of negative real numbers would be West; the upward line of positive imaginaries would be North; and the downward line of negative imaginaries would be South.

Now if we agree that $+1$ times $+1$ equals $+1$, and if we concentrate on the compass signs as I have defined them, we are saying that East times East equals East. Again since -1 times -1 also equals $+1$, West times West equals East. Then, since $+i$ times $+i$ equals -1, and so does $-i$ times $-i$, then North times North equals West and so does South times South.

We can also make other combinations such as -1 times $+i$, which equals $-i$ (since positive times negative yields a negative product even when imaginaries are involved), so that West times North equals South. If we list all the possible combinations as compass points, abbreviating those points by initial letters, we can set up the following system:

$$
\begin{array}{llll}
E \times E = E & E \times S = S & E \times W = W & E \times N = N \\
S \times E = S & S \times S = W & S \times W = N & S \times N = E \\
W \times E = W & W \times S = N & W \times W = E & W \times N = S \\
N \times E = N & N \times S = E & N \times W = S & N \times N = W
\end{array}
$$

There is a very orderly pattern here. Any compass point multiplied by East is left unchanged, so that East as a multiplier represents a rotation of 0°. On the other hand, any compass point multiplied by West is rotated through 180° ("about face"). North and South represent right-angle turns. Multiplication by South results in a 90° clockwise turn ("right face"); while multiplication by North results in a 90° counterclockwise turn ("left face").

Now it so happens that an unchanging direction is the simplest arrangement, so East (the positive real numbers) is easier to handle and more comforting to the soul than any of the others. West (the negative real numbers), which produces an about face but leaves one on the same line at least, is less comforting, but not too bad. North

76

and South (the imaginary numbers), which send you off in a new direction altogether, are least comfortable.

But viewed as compass points, you can see that no set of numbers is more "imaginary" or, for that matter, more "real" than any other.

Now consider how useful the existence of two number axes can be. As long as we deal with the real numbers only, we can move along the real-number axis, backward and forward, one-dimensionally. The same would be true if we used only the imaginary-number axis.

Using both, we can define a point as so far right or left on the real-number axis and so far up or down on the imaginary-number axis. This will place the point somewhere in one of the quadrants formed by the two axes. This is precisely the manner in which points are located on the earth's surface by means of latitude and longitude.

We can speak of a number such as $+5 +5i$, which would represent the point reached when you marked off 5 units East followed by 5 units North. Or you can have $-7 + 6i$ or $+0.5432 - 9.115i$ or $+\sqrt{2} + \sqrt{3}\,i$.

Such numbers, combining real and imaginary units, are called "complex numbers."

Using both axes, any point in a plane (and not merely on a line) can be made to correspond to one and only one complex number. Again every conceivable complex number can be made to correspond to one and only one point on a plane.

In fact, the real numbers themselves are only special cases of the complex numbers, and so, for that matter, are the imaginary numbers. If you represent complex numbers as all numbers of the form $+a +bi$, then the real numbers are all those complex numbers in which b happens to be equal to zero. And imaginary numbers are all the complex numbers in which a happens to be equal to zero.

The use of the plane of complex numbers, instead of the lines of real numbers only, has been of inestimable use to the mathematician.

For instance, the number of solutions in a polynomial equation is equal to its degree only if complex numbers are considered as solutions, rather than merely real numbers and imaginary numbers. For instance the two solutions of $x^2 - 1 = 0$ are $+1$ and -1, which can be written as $+1 +0i$ and $-1 +0i$. The two solutions of $x^2 + 1 = 0$ are $+i$ and $-i$, or $0 + i$ and $0 - i$. The four solutions of $x^4 - 1 = 0$ are all four complex numbers just listed.

In all these very simple cases, the complex numbers contain zeros and boil down to either real numbers or to imaginary numbers. This, nevertheless, is not always so. In the equation $x^3 - 1 = 0$ one solution, to be sure, is $+1 + 0i$ (which can be written simply as $+1$), but the other two solutions are $-\frac{1}{2} + \frac{1}{2}\sqrt{3}\,i$ and $-\frac{1}{2} - \frac{1}{2}\sqrt{3}\,i$.

The Gentle Reader with ambition can take the cube of either of these expressions (if he remembers how to multiply polynomials algebraically) and satisfy himself that it will come out $+1$.

Complex numbers are of practical importance too. Many familiar measurements involve "scalar quantities" which differ only in magnitude. One volume is greater or less than another; one weight is greater or less than another; one density is greater or less than another. For that matter, one debt is greater or less than another. For all such measurements, the real numbers, either positive or negative, suffice.

However, there are also "vector quantities" which possess both magnitude and direction. A velocity may differ from another velocity not only in being greater or less, but in being in another direction. This holds true for forces, accelerations, and so on.

For such vector quantities, complex numbers are necessary to the mathematical treatment, since complex numbers include both magnitude and direction (which was my reason for making the analogy between the four types of numbers and the compass points).

Now, when my sociology professor demanded "the square root of minus one pieces of chalk," he was speaking of a scalar phenomenon for which the real numbers were sufficient.

On the other hand, had he asked me how to get from his room to a certain spot on the campus, he would probably have been angered if I had said, "Go two hundred yards." He would have asked, with asperity, "In which direction?"

Now, you see, he would have been dealing with a vector quantity for which the real numbers are insufficient. I could satisfy him by saying "Go two hundred yards northeast," which is equivalent to saying "Go $100\sqrt{2}$ plus $100\sqrt{2}\ i$ yards."

Surely it is as ridiculous to consider the square root of minus one "imaginary" because you can't use it to count pieces of chalk as to consider the number 200 as "imaginary" because by itself it cannot express the location of one point with reference to another.

7. Pre-fixing It Up

I GO THROUGH LIFE supported and bolstered by many comforting myths, as do all of us. One of my own particularly cherished articles of faith is that there are no arguments against the metric system and that the common units make up an indefensible farrago of nonsense that we keep out of stubborn folly.

Imagine the sobering effect, then, of having recently come across a letter by a British gentleman who bitterly denounced the metric system as being artificial, sterile, and not geared to human needs. For instance, he said (and I don't quote exactly), if one wants to drink beer, a pint of beer is the thing. A liter of beer is too much and half a liter is too little, but a pint, ah, that's just right.[1]

As far as I can tell, the gentleman was serious in his provincialism, and in considering that that to which he is accustomed has the force of a natural law. It reminds me of the pious woman who set her face firmly against all foreign languages by holding up her Bible and saying, "If the English language was good enough for the prophet Isaiah, and the apostle Paul, it is good enough for me."

But mainly it reminds me that I want to write an essay on the metric system.

[1]Before you write to tell me that half a liter is larger than a pint, let me explain that though it is larger than an American pint, it is smaller than a British pint.

In order to do so, I want to begin by explaining that the value of the system does not lie in the actual size of the basic units. Its worth is this: that it is a logical *system*. The units are sensibly interrelated.

All other sets of measurements with which I am acquainted use separate names for each unit involving a particular type of quantity. In distance, we ourselves have miles, feet, inches, rods, furlongs, and so on. In volume, we have pecks, bushels, pints, drams. In weight, we have ounces, pounds, tons, grains. It is like the Eskimos, who are supposed to have I don't know how many dozens of words for snow, a different word for it when it is falling or when it is lying there, when it is loose or packed, wet or dry, new-fallen or old-fallen, and so on.

We ourselves see the advantage in using adjective-noun combinations. We then have the noun as a general term for all kinds of snow and the adjective describing the specific variety: wet snow, dry snow, hard snow, soft snow, and so on. What's the advantage? First, we see a generalization we did not see before. Second, we can use the same adjectives for other nouns, so that we can have hard rock, hard bread, hard heart, and consequently see a new generalization, that of hardness.

The metric system is the only system of measurement which, to my knowledge, has advanced to this stage.

Begin with an arbitrary measure of length, the meter (from the Latin *metrum* or the Greek *metron,* both meaning "to measure"). Leave that as the generic term for length, so that all units of length are meters. Differentiate one unit of length from another by means of an adjective. That, in my opinion, would be fixing it up right.

To be sure, the adjectives in the metric system (lest they get lost by accident, I suppose) are firmly jointed to the generic word and thus become prefixes. (Yes, Gentle Reader, in doing this to the measurement system, they were "pre-fixing it up.")

The prefixes were obtained out of Greek and Latin in accordance with the following little table:

English	Greek	Latin
thousand	chilioi	mille
hundred	hecaton	centum
ten	deka	decem

Now, if we save the Greek for the large units and the Latin for the small ones, we have:

1 kilometer[2]	equals	1000	meters
1 hectometer	equals	100	meters
1 dekameter	equals	10	meters
1 meter	equals	1	meter
1 decimeter	equals	0.1	meter
1 centimeter	equals	0.01	meter
1 millimeter	equals	0.001	meter

It doesn't matter how long a meter is; all the other units of length are as defined. If you happen to know the length of the meter in terms of yards or of wavelengths of light or of two marks on a stick, you automatically know the lengths of all the other units. Furthermore, by having all the sub-units vary by powers of ten, it becomes very easy (given our decimal number system) to convert one into another. For instance, I can tell you right off that there are exactly one million millimeters in a kilometer. Now you tell me right off how many inches there are in a mile.

And again, once you have the prefixes memorized, they will do for *any* type of measurement. If you are told that a "poise" is a measure of viscosity, it doesn't matter how large a unit it is or how it is related to other sorts of units or even what, exactly, viscosity is. Without knowing anything at all about it, you still know that a centipoise is equal to a hundredth of a poise, that a hectare is a hundred acres, that a decibel is a tenth of a bel; and even that a "kilobuck" is equal to a thousand dollars.[3]

In one respect and, to my mind, in only one were the French scientists who established the metric system in 1795 shortsighted. They did not go past the thousand mark in their prefix system.

Perhaps they felt that once a convenient basic unit was

[2]The Greek *ch* has the guttural German *ch* sound. The French, who invented the metric system, have no such sound in their language and used *k* instead as the nearest approach. That is why *chilioi* becomes *kilo*. Since we don't have the guttural *ch* either, this suits us fine.

[3]If anyone wants to write that a millipede is a thousandth of a pede and that one centipede equals ten millipedes, by all means, do—but I won't listen.

selected for some measurable quantity, then a sub-unit a thousand times larger would be the largest useful one, while a sub-unit a thousandth as large would be the smallest. Or perhaps they were influenced by the fact that there is no single word in Latin for any number higher than a thousand. (Words like *million* and *billion* were invented in the late middle ages and in early modern times.)

The later Greeks, to be sure, used *myrias* for ten thousand, so it is possible to say "myriameter" for ten thousand meters, but this is hardly ever used. People say "ten kilometers" instead.

The net result, then, is that the metric system as organized originally offers prefixes that cover only six orders of magnitude. The largest unit, "kilo," is one million (10^6) times as great as the smallest unit "milli," and it is the exponent, 6, that marks the orders of magnitude.

Scientists could not, however, stand still for this. Six orders of magnitude may do for everyday life, but as the advance of instrumentation carried science into the very large and very small in almost every field of measurement, the system simply had to stretch.

Unofficial prefixes came into use for units above the kilo and below the milli and of course that meant the danger of nonconformity (which is a bad thing in scientific language). For instance, what we call a "Bev" (billion electron-volts), the British call a "Gev" (giga-electron-volts).

In 1958, then, an extended set of prefixes, at intervals of three orders of magnitude, was agreed upon by the International Committee on Weights and Measures at Paris. Here they are, with a couple of the older ones thrown in for continuity:

Size	Prefix	Greek Root
trillion (10^{12})	tera-	teras ("monster")
billion (10^9)	giga-	gigas ("giant")
million (10^6)	mega-	megas ("great")
thousand (10^3)	kilo-	
one (10^0)		
thousandth (10^{-3})	milli-	
millionth (10^{-6})	micro-	mikros ("small")
billionth (10^{-9})	nano-	nanos ("dwarf")
trillionth (10^{-12})	pico-	

The prefix *pico-* does not have a Greek root.

Well, then, we have a "picometer" as a trillionth of a meter, a "nanogram" as a billionth of a gram, a "gigasecond" as a billion seconds, and a "tetradyne" as a trillion dynes. Since the largest unit, the tera, is 10^{24} times the smallest unit, the pico, the metric system now stretches not merely over 6, but over a full 24 orders of magnitude.

In 1962 *femto-* was added for a quadrillionth (10^{-15}) and *atto-* for a quintillionth (10^{-18}). Neither prefix has a Greek root. This extends the metric system over 30 orders of magnitude.

Is this too much? Have we overdone it, perhaps? Well, let's see.

The metric unit of length is the meter. I won't go into the story of how it was fixed at its precise length, but that precise length in terms of familiar units is 1.093611 yards or 39.37 inches.

A kilometer, naturally, is a thousand times that, or 1093.6 yards, which comes out to 0.62137 miles. We won't be far off if we call a kilometer ⅝ of a mile. A mile is sometimes said to equal "twenty city blocks"; that is, the distance between, let us say, 59th Street and 79th Street in Manhattan. If so, a kilometer would represent 12½ city blocks, or the distance from halfway between 66th and 67th streets to 79th Street.

For a megameter we increase matters three orders of magnitude and it is equal to 621.37 miles. This is a convenient unit for planetary measurements. The air distance from Boston, Massachusetts, to San Francisco, California, is just about 4⅓ megameters. The diameter of the earth is 12¾ megameters and the circumference of the earth is about 40 megameters. And finally, the moon is 380 megameters from the earth.

Passing on to the gigameter, we have a unit 621,370 miles long, and this comes in handy for the nearer portions of the solar system. Venus at its closest is 42 gigameters away and Mars can approach us as closely as 58 gigameters. The sun is 145 gigameters from the earth and Jupiter, at its closest, is 640 gigameters distant; at its farthest, 930 gigameters away.

Finally, by stretching to the limit of the newly extended metric system, we have the terameter, equal to 621,370,-000 miles. This will allow us to embrace the entire solar system. The extreme width of Pluto's orbit, for instance, is not quite 12 terameters.

The solar system, however, is just a speck in the Galaxy. For measuring distances to the stars, the two most common units are the light-year and the parsec, and both are outside the metric system. What's more, even the new extension of the system can't reach them. The light-year is the distance that light travels in one year. This is about 5,880,000,000,000 miles or 9450 terameters. The parsec is the distance at which a star would appear to us to have a parallax of one second of arc (*par*allax-*sec*ond, get it), and that is equal to 3.26 light-years, or about 30,000 terameters.

Even these nonmetric units err on the small side. If one were to draw a sphere about the solar system with a radius of one parsec, not a single known star would be found within that sphere. The nearest stars, those of the Alpha Centauri system, are about 1.3 parsecs away. There are only thirty-three stars, out of a hundred billion or so in the Galaxy, closer to our sun than four parsecs, and of these only seven are visible to the naked eye.

There are many stars beyond this—far beyond this. The Galaxy as a whole has a diameter which is, at its longest, 30,000 parsecs. Of course, we might use the metric prefixes and say that the diameter of the Galaxy is 30 kiloparsecs.

But then the Galaxy is only a speck in the entire universe. The nearest extragalactic structures are the Magellanic Clouds, which are 50 kiloparsecs away, while the nearest full-size galaxy to our own is Andromeda, which is 700 kiloparsecs away. And there are hundreds of billions of galaxies beyond at a distance of many megaparsecs.

The farthest galaxies that have been made out have distances estimated at about two billion parsecs, which would mean that the entire visible universe, as of now, has a diameter of about 4 gigaparsecs.

Suppose, now, we consider the units of length in the other direction—toward the very small.

A micrometer is a good unit of length for objects visible under the ordinary optical microscope. The body cells, for instance, average about 4 micrometers in diameter. (A micrometer is often called a "micron.")

Drop down to the nanometer (often called a "millimicron") and it can be conveniently used to measure the wavelengths of visible light. The wavelength of the longest red light is 760 nanometers, while that of the shortest violet light is 380 nanometers. Ultraviolet light has a range of wavelengths from 380 nanometers down to 1 nanometer.

Shrinking the metric system to its tiniest, we have the picometer, or a trillionth of a meter. Individual atoms have diameters of from 100 to 600 picometers. And soft gamma rays have wavelengths of about 1 picometer.

The diameter of subatomic particles and the wavelengths of the hard gamma rays go well below the picometer level, however, reaching something like 1 femtometer.

The full range of lengths encountered by present-day science, from the diameter of the known universe at one extreme, to the diameter of a subatomic particle at the other, covers a range of 41 orders of magnitude. In other words, it would take 10^{41} protons laid side by side to stretch across the known universe.

What about mass?

The fundamental unit of mass in the metric system is the gram, a word derived from the Greek *gramma*, meaning a letter of the alphabet.[4] It is a small unit of weight, equivalent to $\frac{1}{2\,8.3\,5}$ ounces. A kilogram, or a thousand grams, is equal to 2.205 pounds, and a megagram is therefore equal to 2205 pounds.

The megagram is almost equal to the long ton (2240 pounds) in our own units, so it is sometimes called the "metric ton" or the "tonne." The latter gives it the French spelling, but doesn't do much in the way of differentiating the pronunciation, so I prefer metric ton.

A gigagram is 100 metric tons and a teragram is 1,000,-000 metric tons and this is large enough by commercial standards. These don't even begin, however, to scratch the surface astronomically. Even a comparatively small body like the moon has a mass equal to 73 trillion teragrams. The earth is 81 times more massive and has a mass of nearly 6 quadrillion teragrams. And the sun, a merely average star, has a mass 330,000 times that of the earth.

Of course, we might use the sun itself as a unit of weight.

[4] The Greeks marked small weights with letters of the alphabet to indicate their weight, for they used letters to represent numbers, too.

For instance, the Galaxy has a total mass equal to 150,-000,000,000 times that of the sun, and we could therefore say that the mass of the Galaxy is equal to 150 gigasuns. Since it is also estimated that in the known universe there are *at least* 100,000,000,000 galaxies, then, assuming ours to be of average mass, that would mean a minimum total mass of the universe equal to 15,000,000,000 terasuns or 100 gigagalaxies.

Suppose, now, we work in the other direction.

A milligram, or a thousandth of a gram, represents a quantity of matter easily visible to the naked eye. A drop of water would weigh about 50 milligrams.

Drop to a microgram, or a millionth of a gram, and we are in the microscopic range. An amoeba would weigh in the neighborhood of five micrograms.

The cells of our body are considerably smaller and for them we drop down to the nanogram, or a billionth of a gram. The average liver cell has a weight of about two nanograms.

Below the cells are the viruses, but even if we drop to the picogram, a trillionth of a gram, we do not reach that realm. The tobacco-mosaic virus, for instance, weighs only 66 attograms.

Nor is that particularly near the bottom of the scale. There are molecules far smaller than the smallest virus, and the atoms that make up the molecules and the particles that make up the atom. Consider the following table:

Weight in Attograms

hemoglobin molecule	0.1
uranium atom	0.0004
proton	0.00000166
electron	0.0000000009

All told, the range in mass from the electron to the minimum total mass of the known universe covers 83 orders of magnitude. In other words, it would take 10^{83} electrons to make a heap as massive as the total known universe.

In some ways, time (the third of the types of measurement I am considering) possesses the most familiar units, because that is the one place where the metric system introduced no modification at all. We still have the second, the minute, the hour, the day, the year, and so on.

This means, too, that the units of time are the only ones used by scientists that lack a systematic prefix system. The result is that you cannot tell, offhand, the number of seconds in a week or the number of minutes in a year or the number of days in fifteen years. Neither can scientists.

The fundamental unit of time is the second and we could, if we wished, build the metric prefixes on those as follows:

1 second	equals	1	second
1 kilosecond	equals	16⅔	minutes
1 megasecond	equals	11⅔	days
1 gigasecond	equals	32	years
1 terasecond	equals	32,000	years

It is sobering to think that I have lived only a little over 1¼ gigaseconds; that civilization has existed for at most about 250 gigaseconds; and that man-like creatures may not have existed for more than 18 teraseconds altogether. Still, that doesn't make much of an inroad into geologic time and even less of an inroad into astronomic time.

The solar system has been in existence for about 150,000 teraseconds and may well remain in existence without major change for 500,000 additional teraseconds. The smaller the star, the more carefully it hoards its fuel supply and a red dwarf may last without undue change for as long as 3,000,000 teraseconds. As for the total age of the universe, past and future, I say nothing. There is no way of estimating, and the continuous-creation boys consider its lifetime to be eternal.

I have one suggestion to make for astronomic time, however (a suggestion which I don't think is particularly original with me). The sun, according to reasonable estimates, revolves about the galactic center once every 200,000,000 years. This we could call a "galactic year" or, better, a "galyear." (An ugly word, but never mind!) One galyear is equal to 6250 teraseconds. On the other hand, a "pico-galyear" is equal to 1 hour and 45 minutes.

If we stick to galyears then, the entire fossil record covers at most only 3 galyears; the total life of the solar system thus far is only 25 galyears; and the total life of a red dwarf as a red dwarf is perhaps 500 galyears.

But now I've got to try the other direction, too, and see what happens for small units of time. Here at least there are no common units to confuse us. Scientists have therefore been able to use *millisecond* and *microsecond* freely, and now they can join to that *nanosecond*, *picosecond*, *femtosecond*, and *attosecond*.

These small units of time aren't very useful in the macroscopic world. When a Gagarin or a Glenn circles the earth at 5 miles a second, he travels less than 9 yards in a millisecond and less than a third of an inch in microseconds. The earth itself, moving at a velocity of 18½ miles a second in its travels about the sun, moves only a little over an inch in a microsecond.

In other words, at the microsecond level, ordinary motion is frozen out. However, the motion of light is more rapid than any ordinary motion, while the motion of some speeding subatomic particles is nearly as rapid as that of light. Therefore, let's consider the smaller units of time in terms of light.

Distance Covered by Light

1 second	186,200	miles
1 millisecond	186	miles
1 microsecond	327	yards
1 nanosecond	1	foot
1 picosecond	$\frac{1}{80}$	inch

Now, you may think that at picosecond levels subatomic motion and even light-propagation is "frozen." After all, I dismissed earth's motion as "frozen" when it moved an inch. How much more so, then, when thousandths of an inch are in question.

However, there is a difference. The earth, in moving an inch, moves $\frac{1}{500,000,000}$ of its own diameter. A speeding subatomic particle moving at almost the speed of light for a distance of $\frac{1}{80}$ of an inch moves 120,000,000,000 times its own diameter. To travel a hundred and twenty billion times its own diameter, the earth would have to keep on going for 1,500,000 years. For Gagarin or Glenn to have traveled for a hundred and twenty billion times their own diameter, they would have had to stay in orbit a full year.

A subatomic particle traveling $\frac{1}{80}$ of an inch is therefore anything but "frozen," and has time to make a fabulous number of collisions with other subatomic particles or to undergo internal changes. As an example, neutral pions break down in a matter of 0.1 femtoseconds after formation.

What's more, the omega-meson breaks down in something like 0.0001 attoseconds or, roughly, the time it would take light to cross the diameter of an atomic nucleus and back.

The entire range of time, then, from the lifetime of an omega-meson to that of a red-dwarf star covers a range of 40 orders of magnitude. In other words, during the normal life of a red dwarf, some 10^{40} omega-mesons have time to come into existence and break down, one after the other.

To summarize, the measurable lengths cover a range of 41 orders of magnitude, the measurable masses 83 orders of magnitude, and the measurable times 40 orders of magnitude. Clearly, we are not overdoing it in expanding the metric system from 6 to 30 orders of magnitude.

Part Two

PHYSICS

8. The Rigid Vacuum

PROBABLY THE GREATEST dilemma facing the man who wants to write science fiction on the grand scale—and who is also conscientious—is that of squaring the existence of an interstellar society with the fact that travel at velocities greater than that of light in a vacuum (186,200 miles per second) is considered impossible.

There are a number of ways out, however, and I'll mention three. The most honest is to accept the limitation, and to assume instead that travelers experience time-dilatation. That is, a trip that takes two weeks from their own standpoint may take twenty years from the standpoint of those at home. This, of course, creates difficulties of plotting, and is therefore unpopular among most writers.

The most daring and intriguing solution is E. E. Smith's "inertialess drive," in which matter is assumed to be freed of inertia. (As far as we know, by the way, this is impossible.) Matter without inertia can undergo an acceleration of any size by the application of any force however small. Smith assumed that matter would then be capable of attaining any velocity, even one far beyond that of light.

Actually, this is unlikely. Photons and neutrinos have zero mass and therefore zero inertia, yet travel no faster than the velocity of light. Consequently, an inertialess drive would be of no help.

There's another flaw here, too. The resistance of even the thinly-spread gas and dust in interstellar space would become significant as velocity rises. Eventually a limit to velocity would be set beyond which one could expect the ship to be melted and its occupants broiled.

The most pedestrian solution is the one I use myself, which is to speak of "hyperspace." This involves higher dimensions and one usually drags in analogies concerning one's going through a piece of paper to get to the other side, instead of traveling all the way over to the far distant edge.

In contrast to all this great thought given over to the problem of interstellar travel, very little is devoted to the problem of interstellar communication. According to the relativistic viewpoint of the universe, it is not simply matter that cannot be transported at speeds greater than that of light in a vacuum; it is any form of meaningful symbol.

Well, then, suppose you don't want to travel to Sirius to see your girl friend; suppose you just want to put in a call and speak to her. How do you do that without having to wait sixteen years for the signal to make the round trip.

As far as I know, when this facet of the problem is considered, it is tossed off with the word *sub-etheric*.[1] And that, at last, brings me to the point. I want to explain what a science-fiction writer means by *sub-etheric*, and I want to do it in my own fashion; i.e., the long way round.

The word *ether* has had a long and splendid history, dating back to the time it was coined by Aristotle about 350 B.C.

To Aristotle the manner in which an object moved was dictated by its own nature. Earthy materials fell and fiery particles rose because earthy materials had an innate tendency to fall and fiery particles an innate tendency to rise. Therefore, since the objects in the heavens seemed to move in a fashion characteristic of themselves (they moved circularly, round and round, instead of vertically, up or down), they had to be made of a substance completely different from any with which we are acquainted down here.

It was impossible to reach the heavens and study this mysterious substance, but it could at least be given a name. (The Greeks were good at making up names, whence the

[1] At least, that's how *I* toss it off.

phrase, "The Greeks had a word for it.") The one property of the heavenly objects that could be perceived, aside from their peculiar motion, was, however, their blazing luminosity. The sun, moon, planets, stars, comets, and meteors all gave off light. The Greek word for "to blaze" (transliterated into our alphabet) is *aithein*. Aristotle therefore called the heavenly material *aither*, signifying "that which blazes." In Aristotle's day it was pronounced "i'ther," with a long *i*.

The Romans adopted this Greek word, because to the Romans, Greek was the language of learning and the average Roman pedant adapted all the Greek words he could, just as our modern pedants are as Latinized as possible, and as the pedant of the future will drag in all the ancient English he can. The Romans transliterated *aither* into *aether*, making use of the diphthong *ae* to keep the pronunciation correct, since that, in the Latin of Cicero's day, was pronounced like a long *i*. (*Caesar* is pronounced "Kaiser," as the Germans know, but we don't.)

The British keep the Latin spelling of *aether*, but Latin (and Greek, too) underwent changes in pronunciation after classical times, and by medieval times *ae* had something of a long *e* sound. So *aether* came to be pronounced "ee'ther."

But if it's going to be pronounced that way, why not get rid of the superfluous *a* and spell it "ether." This, actually, is what Americans do.

(The Greek word for blood is *haima*, and now you can figure out for yourself why we write "hemoglobin" and the British write "haemoglobin.")

This Aristotelian sense of the word *ether* is still with us whenever we speak of something that is heavenly, impalpable, refined of all crass material attributes, incredibly delicate, and so on and so on, as being "ethereal."

By 1700 the Greek scheme of the universe had fallen to pieces. The sun, not the earth, was the center of the planetary system, and the earth moved about the sun, as did the other planets. The motions of the heavenly bodies, including the earth, were dictated solely by gravity; and the force of gravity operated on ordinary objects as well. The laws of motion were the same for all matter and did not in the least depend on the nature of the moving object.

Seeming differences were the result of the intrusion of additional effects: buoyancy, friction, and so on.

In the general smashup of Aristotelian physics, however, one thing remained—the ether.

You see, if we wipe out the notion that objects move according to some inner compulsion, then they must move according to some compulsion imposed upon them from outside. This outer compulsion, gravity, bound the earth to the sun, for instance—but, come to think of it, how?

If you wish to exert a force on something; to push it or pull it; you must make contact with it. If you do not make direct contact with it, then you make indirect contact with it; you push it with a stick you hold in your hand or pull it with a hook. Or you can throw a stick (or a boomerang) and the force you impart to the stick is carried, physically, to the object you wish to affect. Even if you knock down a house of cards with a distant wave of the hand, it is still the air you (so to speak) throw at the cards, that physically carries the force to the cards.

In short, something physical must connect the object forcing and the object forced. Failing that, you have "action at a distance," which is a hard thing to grasp and which philosophers of science seem to be reluctant to accept if they can think of any other way out of a dilemma.

But gravity seems to involve action at a distance. Between the sun and the earth, or between the earth and the moon, is a long stretch of nothing, not even air. The force of gravitation makes itself felt across the vacuum; it is therefore conducted across it; and the question arises: What does the conducting? What carries the force from the sun to the earth?

The answer consisted of Aristotle's word again, *ether*. This new ether, however, was not something that made up the heavenly bodies. The seventeenth-century scientist rather suspected the heavenly bodies were made up of ordinary earthly matter. Instead, ether was now viewed as making up the apparently empty volume through which all these bodies of matter moved. In short, it made up space; it was, so to speak, the very fabric of space.

Exactly what ether's properties were could not be shown by direct observation, for it could not be directly observed. It was not matter or energy, for when only ether was present, what seemed to exist to our senses and to our measurements was a vacuum—*nothing*. At the same time, ether (whatever it was) was to be found not only in empty space but permeating all matter, too, for the conduction of the gravitational force did not seem to be interfered with by matter. If, as during a solar eclipse, the moon passed between the earth and the sun, the earth's movements were not affected by a hair. The force of gravity clearly traveled, unchanged and undiminished, through two thousand miles of matter. Consequently, the ether permeated the moon and, by a reasonable generalization, it permeated all matter.

Furthermore, ether did not interfere with the motion of the planets. Planets moved through the ether as though it were not there. Matter and ether, then, simply did not interact at all. Ether could conduct forces but was not itself subject to them.

This meant that ether was not moving. How could it move unless some force were applied to it, and how could such a force be exerted upon it if matter would not interact with it? Or, to put it another way, ether is indistinguishable from a vacuum, and can you picture a way in which you can exert force on a vacuum (not on a container which may hold a vacuum, but on the vacuum itself) so as to impart motion to it?

This was an important point. As long as astronomers were sure that the earth was the motionless center of the universe (even if it rotated, the *center* of the earth was motionless), it was possible to work up laws of motion with confidence. Motion was a concept that meant something. If the earth traveled about the sun, however, then while you were working out the laws of motion relative to the earth, you would be plagued by wondering whether those laws would make sense if the same motion were viewed relative to Mars, for instance.

Actually, if one could find something that was at rest and refer motion to that, then the laws of motion would still make sense because the earth's motion with reference to the something at rest could be subtracted from the object's motion with reference to the something at rest and that would leave the object's motion with reference to the earth, and the laws would still apply and you wouldn't have to worry about the motions with reference to Mars or to Alpha Centauri or anything else.

And this was where the ether came in. Ether could not

move; motion was alien to the very concept of ether; so it could be considered in a state of Absolute Rest. This meant there was such a thing as Absolute Motion, since any motion could, in principle, be referred to the ether. The framework of space and time within which such absoluteness of rest and motion can exist can be referred to as Absolute Space and Absolute Time.

A century after Newton, ether was to be called upon again. The force of gravity, after all, was not the only entity to reach us across the stretches of empty space; another entity was light.

Light did not, however, raise the anxiety at first that gravitation did, for it did not act as gravity did. For one thing, light could be shielded. When the moon interposed itself between ourselves and the sun, light was cut off even though gravity wasn't. Thin layers of matter could completely block even strong light, so that it would seem that light could not be conducted by the ether which permeated all matter.

Furthermore, the direction in which a light ray traveled could be changed ("refracted") by passing it from one medium to another, as from air to water, although ether permeated both media equally. The direction in which gravity exerted its force could not be changed by any known method.

Newton postulated, therefore, that light consisted of tiny particles moving at great velocities. In this way, light required no ether and yet did not represent action at a distance either, for the effect was carried across a vacuum, physically, by moving objects. Furthermore, the particle theory could be easily elaborated to explain the straight-line motion of light, and its ability to be reflected and refracted.

There were opposing views in Newton's time to the effect that light was a wave form, but this made no headway. The wave forms then known (water waves and sound waves, for instance) did not travel in straight lines but easily bent around obstacles. This was not at all the way light acted and therefore light could not be a wave form.

In 1801, however, an English physician, Thomas Young, showed that it was possible to combine two rays of light in such a way as to get alternating bands of light and darkness ("interference fringes"). This seemed difficult to explain if light consisted of particles (for how could two particles add together to make no particles?), but very easy to explain if light were a wave form. Suppose the wave of one light ray were on its way up and the wave of the other were on its way down. The two effects would cancel, for no net motion at all, and there would be darkness.

Furthermore, it could be shown that a wave form would move about obstacles that were of a size comparable to its own wavelength. Obstacles larger than that would be increasingly efficient (as their size increased) in reflecting the wave form. Where obstacles vastly larger than the wavelength were concerned, the wave form would seem to travel in straight lines and cast sharp shadows.

Well, ordinary sound waves have wavelengths measured in feet and yards. Young, however, was able to deduce the wavelength of light from the width of the interference fringes and found it to be something like a sixty-thousandth of an inch. As far as obstacles of ordinary size were concerned, obstacles large enough to see, light traveled in straight lines and cast sharp shadows even though it was a wave form.

But this new view did not take over without opposition. It raised serious philosophical problems. It makes one ask at once: "If light consists of waves, then what is waving?" In the case of water waves, water molecules are moving up and down. In the case of sound waves through air, air molecules are moving to and fro. But light waves?

The answer was forced upon physicists. Light can travel through a vacuum with the greatest ease, and the vacuum contained nothing but ether. If light was a wave form, it had to consist, therefore, of waves of ether.

But then how account for the fact that light could be reflected, refracted, and absorbed, when gravitation carried by the same ether could not? Was it possible that there were two ethers with different properties, one to conduct gravity and one to conduct light? The question was never answered, but through the nineteenth century, light was far more crucial to the development of theoretical physics than gravity was, and it was the particular ether that carried light that was under continual discussion. Physicists referred to it as the "luminiferous ether" (Latin for "light-carrying ether").

But difficulties were to arise in the case of the luminiferous ether that never arose in the case of the gravity-carrying ether. You see, there are two kinds of wave forms—

In water waves, while the wave motion itself is progressing, let us say, from right to left, the individual water molecules are moving up and down. The movement of the oscillating parts is in a direction at right angles to the movement of the wave itself. This type of wave form, resembling a wriggling snake, is a "transverse wave."

In sound waves the individual molecules are moving back and forth in the same direction that the sound wave is traveling. Such a wave form (a bit harder to picture) is a "longitudinal wave."

Well, then, what kind of a wave is a light wave, transverse or longitudinal? At first, everyone voted for longitudinal waves—even Young did—for reasons I'll shortly explain.

Unfortunately, one annoying fact intervened. Back in Newton's time a Dutch physician, Erasmus Bartholin, had discovered that a ray of light, upon entering a transparent crystal of a mineral called Iceland spar, was split into two rays. The separation was brought about because the original ray was bent by two different amounts. Everything seen through Iceland spar seemed double, and the phenomenon was called "double refraction."

In order for a ray of light to bend in two different directions on entering Iceland spar, the components of light had to exist in two different varieties, or, if there were only one variety, that variety had to show some sort of asymmetry.

Newton tried to adjust the particle theory of light to account for this, and made a heroic effort, too. Through sheer intuition, he caught a glimmer of our modern view of light as consisting of both particles and waves, two centuries ahead of time. However, after Newton's death, the lesser minds that followed him thought of a much better way of accounting for "double refraction." They ignored it.

What about the wave theory? Well, no one could think of a way to make a longitudinal wave explain double refraction, but transverse waves were another matter.

Imagine that your eye is a piece of Iceland spar and that a ray of light is coming directly toward it. The ether, as was then supposed, would be undulating at right angles to

97

the direction of motion, but there are an infinite number of directions that would be at right angles to the direction of motion. As the light comes toward you, the ether could be moving up and down, or right and left, or diagonally (turned either clockwise or counterclockwise), to any extent.

Every diagonal undulation can be divided into two components, a vertical one and a horizontal one, so in the last analysis we can say that the light ray approaching us is made up of vertical undulations and horizontal undulations. Well, Iceland spar can choose between them. The vertical undulations bend to one extent, the horizontal to another, and where one ray of light enters, two emerge.

It is a good question as to why Iceland spar should do this and not glass, but the question is not pertinent to this discussion and I shall leave that to another essay some day. What does matter is simply that longitudinal waves could not be used to explain double refraction and transverse waves could and the conclusion had to be, then, that light consisted of transverse waves. The theory of light as a transverse wave form was worked out in the 1820s by a French physicist named Augustin Jean Fresnel.

This aroused a furor indeed, for the manner in which longitudinal waves and transverse waves are conducted show important differences. Longitudinal waves can be conducted by matter in any state, gaseous, liquid, or solid. Thus, sound waves travel through air, through water, and through iron with equal ease. If light were a longitudinal wave, then the luminiferous ether could be viewed as an exceedingly subtle gas; so subtle as to be indistinguishable from a vacuum. It would still be capable, in principle, of conducting light.

Transverse waves are more particular. They cannot travel through the body of a gas or liquid. (Water waves agitate the surface of water, but cannot travel through the water itself.) Transverse waves can travel through solids only. This means that if the luminiferous ether conducts light, and if light is a transverse wave, then the luminiferous ether must have the properties of a solid!

And there is worse to follow. For atoms or molecules to engage in periodic motion (as they must, to establish a wave form), they must have elasticity. They must spring back into position, if deformed out of it, overshoot the

98

mark, spring back again, overshoot the mark again, and so on. The speed with which an atom or molecule springs back into position depends upon the rigidity of the material. The more rigid, the faster the snapback, the faster the oscillation as a whole, and the faster the progress of the wave form. Thus sound waves progress more rapidly through water than through air, and more rapidly through steel than through water.

It works in reverse. If we know the velocity at which a wave form travels through a medium, we can calculate how rigid it must be.

Well, what is the velocity of light through a vacuum; i.e., through the ether? It is 186,200 miles per second, and this was known in Fresnel's time. For transverse waves to travel that rapidly, the conducting medium must be rigid indeed—more rigid than steel.

And so there's the picture of the luminiferous ether; a substance indistinguishable from a vacuum yet more rigid than steel. A rigid vacuum! No wonder physicists tore their hair.

A generation of mathematicians worked out theories to account for this wedding of the mutually exclusive and managed to cover the general inconceivability of a rigid vacuum with a glistening layer of fast-talking plausibility. As for an actual physical picture of the luminiferous ether, the best that could be advanced was that it was a substance something like the modern Silly Putty. It yielded freely to a stress applied relatively slowly (as by a planet moving at two to twenty miles per second), but rigidly resisted a stress applied rapidly (as by light traveling at 186,200 miles a second).

Even so, physicists would undoubtedly have given up the ether in despair if it weren't so useful as the only way to avoid action at a distance. And instead of growing less useful with time, it grew more so, thanks to the work of the Scottish mathematician James Clerk Maxwell. This came about as follows.

Long before Newton had worked out the theory of gravitation, two other types of action-at-a-distance forces were known: magnetism and static electricity. Both attracted objects even across a vacuum and both types of forces, it therefore seemed, had to be conducted by the ether. (In fact, before the theory of gravitation had been put forth, men such as Galileo and Kepler speculated that magnetic forces must bind planets to the sun.)

But there again—was there a separate ether for magnetism and one for electricity, as well as one for light and one for gravity? Were there four ethers altogether, each with its own properties? If so, things were worse than ever. This piling up of four different vacuums, one as rigid as steel, and the other three who-knows-what, threatened to rear a structure that would topple under its own weight and bury the edifice of physics in its ruins.

In the mid-nineteenth century, Maxwell subjected the matter to acute mathematical analysis and showed that he could build up a consistent picture of what was known of electricity and magnetism, and in so doing, maintained that the two forces were interrelated in such a way that one could not exist without the other. There was neither electricity nor magnetism, but "electromagnetism."

Furthermore, if an electrically charged particle oscillated, it radiated energy in the form of a wave, with a frequency equal to that of the oscillation period. In other words, if the charge oscillated a thousand times a second, a thousand waves were formed each second. The velocity of such a wave worked out to a certain ratio which, once solved, turned out to be just about exactly the speed of light.

Maxwell could not believe this to be a coincidence. Light, he insisted, was an "electromagnetic radiation." (Light has a frequency of several hundred trillion waves per second, and where was the electric charge that oscillated at such a rate? Maxwell couldn't answer that, but a generation later, the electrons within the atom were discovered and the question was answered.)

Such a theory was delightful. It unified electricity, magnetism, and light into different aspects of one phenomenon and made one ether do for all three.[2] This simplified the ether concept and made it explain much more than before. (At this point, it should perhaps have been renamed the "electromagniferous ether," but it wasn't.) If Maxwell's theory held up, physicists could grow much more comfortable with the ether concept. But would it hold up?

---

[2]This leaves gravity out, but all efforts to join gravity to electromagnetism as a fourth aspect (a "unified field theory") have failed. Einstein devoted half his life to it but did not succeed.

One way to establish a theory is to make predictions based upon its tenets and have them turn out to be so. To Maxwell, it seemed that since electric charges could oscillate in any period, there should be a whole family of electromagnetic radiations with frequencies greater than those of light and smaller than those of light, and to all degrees.

This prediction was borne out in 1888 (after Maxwell's too-early death, unfortunately) when German physicist Heinrich Hertz managed to get an electric current to oscillate not very rapidly and then detected very low-frequency electromagnetic radiation. This low-frequency, long-wavelength radiation is what we now call "radio waves."

Radio waves, being electromagnetic radiation, are conducted through the ether at 186,20 miles per second. This is the limiting speed of communication by any form of electromagnetic radiation.

But if we grant the ether concept, suppose we imagine a "sub-ether," one that permeates the ether itself as ether permeates matter, and one that has all the properties of ether greatly intensified. It would be even more tenuous and undetectable and at the same time far more rigid. It would, in other words, be a super-rigid super-vacuum. It might even be conjectured that gravitational force, still unaccounted for by Maxwell's theory, would travel through such a sub-ether.

In that case, wave forms (perhaps gravitic, rather than electromagnetic) would travel through it at far greater velocities than that of light. The stars of the galactic empire might then not be too far apart for rapid communication.

And there is your word *sub-etheric*.

Now isn't that an exciting idea? Might it not even be valid? After all, if the ether concept is granted . . .

Ah, but is it granted?

You see, Hertz's discovery of waves that confirmed Maxwell's electromagnetic theories and seemed to establish the ether concept once and for all, had come too late. Few realized it at the time, but the year before Hertz's discovery, the ether concept had been shattered past retrieval.

It happened through one little experiment that didn't work.

And if you read the next chapter, you'll learn about it.

9. The Light That Failed

IN THE SUMMER OF 1962 a fetching young lady from *Newsweek* asked permission to interview me; permission which I granted at once, you may be sure. It seems that *Newsweek* was planning to do a special issue on the space age, and it was this young lady's job to gather some comments on the matter by various science-fiction personalities.

I discoursed learnedly on science fiction to her, filling ninety unforgiving minutes with sixty seconds' worth of distance run each, before I ungripped her with my glittering eye.

Eventually, the special issue appeared, dated October 8, 1962, and there, on page 104, was three quarters of a page devoted to science fiction (and not bad commentary either; no complaints on that score). Within that section, every bit of my long-winded brilliance was discarded with the exception of one remark which read as follows:

"But sci-fi is 'a topical fairy tale where all scientists' experiments succeed,' comments Isaac Asimov. . . ."[1]

Ever since then, this quotation has bothered me. Oh, I said it; I wasn't misquoted. It's just that I seem to have implied that what scientists want are experiments that succeed, and that is not necessarily true.

Under the proper circumstances, a failure, if unexpected and significant, can do more for the development of science than a hundred routine successes. In fact, the most dramatic single experiment in the last three and a half centuries was an outright failure of so thunderous a nature as to win its perpetrator a Nobel Prize. What a happy fairy tale for scientists it would be if all experiments failed like that!

[1] My natural modesty forbids my quoting the rest of the passage, but you can look it up for yourself if you like. Please do.

This fits in, fortunately, with the fact that in the previous chapter, I discussed the ether and ended at a point where it seemed enthroned immovably in the very fabric of physics. I said then that at the very peak of its power and prosperity, the ether was dashed from its throne and destroyed.

The man who brought about that destruction was an American physicist named Albert Abraham Michelson. What started him on the track was a peculiar scientific monomania; Michelson got his kicks out of measuring the velocity of light. Such a measurement was his first scientific achievement, and his last, and just about everything he did in science in between grew out of his perpetual efforts to improve his measurements.

And if you think I'm going to go one step farther without retreating three centuries to discuss the history of the measurement of the velocity of light, you little know me.

Throughout ancient and medieval times the velocity of light was assumed (by those who thought about the matter at all) to be infinite. The Italian scientist Galileo was the first to question this. About 1630 he proposed a method for measuring the velocity of light.

Two people, he suggested, were to stand on hilltops a mile apart, both carrying shielded lanterns. One was to uncover his lantern. The other upon seeing the light was at once to uncover his own lantern. If the first man measured the time that elapsed between his own uncovering and the sight of the spark of light from the other hill, he would know how long it took light to cover the round distance. Galileo had actually tried the experiment, he said, but had achieved no reasonable results.

It is not hard to see why he had failed, in the light of later knowledge. Light travels so quickly that the time lapse between emission and return was far too short for Galileo to measure with any instrument that then existed. There would be a small time lapse to be sure, but that represented the time it took for the assistant to think "Hey, there's the old man's light" and get his own light uncovered.

All that Galileo could possibly have shown by his experiment, which was correct in principle, was that if the velocity of light was not infinite, then it was at least very, very fast by ordinary standards. Still, it was useful to show even this much.

The next step was taken nearly half a century later. In 1676 a Danish astronomer Olaus Roemer was working at the Paris Observatory, observing Jupiter's satellites. Their times of revolution had been carefully measured, so it seemed possible to predict the exact moments at which each would pass into eclipse behind Jupiter, and this, too, had been done.

To Roemer's surprise, however, the moons were being eclipsed at the wrong times. At those times of the year when the earth was approaching Jupiter, the eclipses came more and more ahead of schedule, while when earth was receding from Jupiter, they fell progressively further behind schedule.

Roemer reasoned that he did not see an eclipse when it took place, but only when the cut-off end of the light beam reached him. The eclipse itself took place at the scheduled moment, but when the earth was closer than average to Jupiter, he *saw* the eclipse sooner than if the earth were farther than average from Jupiter. Earth was at a minimum distance from Jupiter when both planets were in a line on the same side of the sun, and it was at a maximum distance when both planets were in a line on opposite sides of the sun. The difference between those distances was exactly the diameter of the earth's orbit.

The difference in time between the earliest eclipse of the satellites and the latest eclipse must therefore represent the time it took light to travel the diameter of earth's orbit. Roemer measured this time as twenty-two minutes and, accepting the best figure then available for the diameter of the earth's orbit, calculated that light traveled at a velocity of 138,000 miles per second. This is only three quarters of what is now accepted as the correct value,[2] but it hit the correct order of magnitude and for a first attempt that was magnificent.

Roemer announced his results and it made a small splash

---

[2] The actual maximum difference of eclipse times, by later measurements, turned out to be sixteen minutes thirty-six seconds. The diameter of the earth's orbit is about 185,500,000 miles, and I leave it to you, O Gentle Reader, to calculate a good approximation of the correct velocity.

but aroused as much opposition as approval and the matter was forgotten for another half century.

In the 1720s the English astronomer James Bradley was hot on the trail of the parallax of the stars. This had become a prime astronomical problem after Copernicus had first introduced the heliocentric theory of the solar system. If the earth really moved about the sun, said the anti-Copernicans, then the nearby stars should seem to shift position ("parallactic displacement") when compared with the more distant stars. Since no such shift was observed, Copernicus must be wrong.

"Ah," said the Copernicans in rebuttal, "but even the nearest stars are so distant that the parallactic displacement is too small to measure."

Yet even after astronomers had all adopted the heliocentric theory, there was still discomfort over the question of the stellar parallax. This business of "too small to measure" seemed very much like an evasion. Observation should be refined to the point where the shift *could* be measured. That would accomplish two things. It would show how far the nearest stars were, and it would be the final proof that the earth was moving round the sun.

Bradley's close observations did, indeed, demonstrate that the stars showed a tiny displacement through the year. However, this displacement was not of the right sort to be explained by the earth's motion. Something else had to be responsible and it was not until 1728 that a suitable explanation occurred to Bradley.

Suppose we consider the starlight bombarding the earth to be like rain drops falling in a dead calm. If a man were standing motionless in such a rainstorm, he would have to hold an umbrella vertically overhead to ward off the vertically-dropping rain. If he were to walk, however, he would be walking into the rain and he would have to angle the umbrella forward, or some drops that would just miss the umbrella would nevertheless hit him. The faster he walks, the greater the angle at which he would have to tilt his umbrella.

In the same way, to observe light from a moving earth, the telescope has to be angled very slightly. As the earth changes the direction of its motion in its course about the sun, the slight angle of the telescope must be changed constantly and the star seems to mark out a tiny ellipse against the sky (with reference to the sun). Bradley had discovered what is called the "aberration of light."

This was not parallactic displacement and it did not help determine the distance of any stars (that had to wait still another century). Just the same it did prove the earth was moving with respect to the stars, for if the earth were motionless, the telescope would not have to be tilted at all, and the star would not seem to move.

It gave additional information, too. The amount of the aberration of light depended on two factors: the velocity of light and the velocity of the earth's motion in its orbit. The latter was known (about 18½ miles per second); therefore the former could be calculated. Bradley's estimate was that light had a velocity of about 188,500 miles per second. This was only 1.2 percent above the true value.

Two independent astronomical methods had yielded figures for the velocity of light and improved observations showed the two methods yielded roughly the same answer. Was there no way, however, in which the velocity could be measured on earth, under conditions controlled by the experimenter?

The answer was yes, but the world had to wait a century and a quarter after Bradley's discovery before a method was found. The discoverer was a French physicist Armand Hippolyte Louis Fizeau who returned to Galileo's method but eliminated the personal element. Instead of having an assistant return a second beam of light, he had a mirror reflect the first one.

In 1849 Fizeau set up a rapidly turning toothed disk on one hilltop and a mirror on another, five miles away. Light passed through one gap between the teeth of the turning disk and was reflected by the mirror. If the disk turned at the proper speed, the reflected light returned just as the next gap moved into line.

From the velocity at which the wheel had to be turned in order for the returning light to be seen, it was possible to calculate the time it took light to cover the ten-mile-round distance. The value so determined was not as good as those the best astronomic measurements had provided; it was 5 percent off, but it was excellent for a first laboratory attempt.

In 1850 Fizeau's assistant Jean Bernard Leon Foucault improved the method by using two mirrors, one of which revolved rapidly. The revolving mirror reflected the light at an angle and from that angle, the velocity of light could be calculated. By 1862 he had obtained values within a percent or so of the true one.

Foucault went further. He measured the velocity of light through water and other transparent media (you could do this with laboratory methods but not with astronomical methods). He discovered in this way that light moved more slowly in water than in air.

This was important. If the particle theory of light were true, light should move more rapidly in water than in air; if the wave theory of light were true, light should move more slowly in water than in air. By the mid-nineteenth century, to be sure, most physicists had accepted the wave theory. Nevertheless, Foucault's experiment was widely interpreted as having placed the final nail in the coffin of the particle theory.

And now we come to Michelson. Michelson had been born in 1852 in a section of Poland that at that time was under German rule, and he was brought to the United States two years later. His family did not follow the usual pattern of settling in one of the large East Coast cities. Instead, the Michelsons made their way out to the far West, a region which the forty-niners had just ripped wide open.

The Michelson family did well there (as merchants, not as goldminers), and young Albert applied for entrance into Annapolis in 1869. He passed the necessary tests, but the son of a war veteran (Civil War, of course) took precedence. It took the personal intervention of President Grant (with an assist from the Nevada congressman who pointed out the political usefulness of such a gesture to the family of a prominent Jewish merchant of the new West) to get Albert in.

He graduated in 1873 and served as a science instructor at the Academy during the latter part of that decade. In 1878 the velocity-of-light bug bit him and he never recovered. Using Foucault's method but adding some ingenious improvements, he made his first report on the velocity, which he announced as 186,508 miles per second. He was approximately 300 miles per second too high, but his result was within one sixth of 1 percent of the actual figure.

In 1882 he tried again, after some years spent in studying optics in Germany and France. This time he came out with a figure of 186,320 miles per second, which was 120 miles per second high, or only one fifteenth of 1 percent off.

Meanwhile, though, it had occurred to Michelson that speeding light could be made to reveal some fundamental secrets of the universe.

One of the big things in the 1880s was the "luminiferous ether" (see the previous chapter). The ether was considered to be motionless, at "absolute rest," and if light were ether waves, then its velocity, if measured carefully enough under appropriate conditions, could give the value of the absolute velocity of the earth; not just its velocity with respect to the sun, but with respect to the very fabric of the universe. Such a value would be of the utmost importance to the philosophy of science, since without it one could never be sure of the validity of all the laws of mechanics that had been worked out since the time of Galileo.

Let me explain how this works. Suppose an airplane were moving at 150 miles an hour and encountered winds of 145 miles an hour. If it were traveling with the wind, the plane would seem to move at 295 miles an hour (as viewed from the ground). If it traveled against the wind, it would travel only five miles an hour with respect to the ground. If the velocity of the plane on a windless day were known, then from the difference in velocity produced by the wind, the wind's own velocity could be calculated.

Now suppose the earth were moving through the stationary ether. From a mechanical standpoint, this would be equivalent to the earth standing still while the ether moved past it. Let's take the latter view, for an "ether wind" is easy to visualize.

Light, consisting (as was thought) of ether waves, would move—relative to the earth—with the ether, moving faster than average in the direction of the ether wind, more slowly than average against that direction, and at intermediate velocities in intermediate directions.

Clearly, the velocity of the ether wind could not be very great. If it were blowing at a considerable fraction of the velocity of light, then all sorts of strange phenomena would be observable. For instance, light would radiate outward in an egg-shaped curve instead of in a circle. The fact that no such phenomena were ever observed, meant that the effects must be very small and that earth's absolute velocity could only be a small fraction of the velocity of light.

Michelson turned his attention toward the possibility of measuring that small fraction.

In 1881 Michelson had constructed an "interferometer," a device which was designed to split a light beam in two, send the parts along different paths at right angles to each other, then bring them back together again.

The two rays of light were made to travel exactly the same distance in the process of going and returning, and therefore, presumably spent the same time on their travels. On returning to their starting point they would merge into one beam again, just as though they had never separated. The merged beams would then display no properties that the original beam had not had.

If, on the other hand, the two light rays had been on their travels for different times, the wave forms of the two rays of light would no longer match; upon merging they would find themselves out of step. There would be places where the waves of one light ray would be moving up while the waves of the other would be moving down. There would then be mutual cancellation ("interference"), and darkness would be produced. The areas of darkness would recur periodically and take the form of a kind of zebra-stripe arrangement ("interference fringes").

The idea was to adjust the instrument so that, as far as was humanly possible, the two light rays would be made to travel just the same distance, and if they then spent the same time at their travels, they would match on merging and no interference fringes would appear.

However, this did not allow for the ether wind, the existence of which was then assumed. If one of the light rays went with the wind, it would return against the wind. The other light ray, sent out at right angles, would then go cross-wind and return cross-wind. It can be shown that, in this case, the time taken by the one ray of light to travel with the wind, then return against it, is slightly longer than the time taken by the other ray of light to travel cross-wind both ways.

The stronger the ether wind, the greater the discrepancy in time; and the greater the discrepancy in time, the wider the interference fringes. By observing the interference fringes, then, Michelson would be able to measure the velocity of the ether wind, and that would give the earth's absolute motion.

Michelson tried the experiment first in Germany, anchoring his interferometer to rock, and driving himself mad trying to eliminate the vibrations set up by city traffic. When he finally sent his split beams of light on their separate paths, he found they brought back no information. The light had failed, and failed miserably. It brought back nothing; no interference fringes at all.

Something had gone wrong, but Michelson did not know what. He let the matter go for a few years.

He returned to the United States, resigned from the Navy, joined the faculty of a new school called Case School of Applied Science, in Cleveland, and there met a chemist named Edward Williams Morley. Morley's ambition had been to be a minister and he took his chemist's job only on condition that he could preach in the school's chapel. His own piece of scientific monomania lay in comparing the atomic weights of oxygen and hydrogen.

Michelson and Morley discussed the interferometer experiment and finally, in 1886, joined forces in order to try again under conditions of the most heroic precautions. They dug down to bedrock to anchor the equipment to the solid planet itself. They built a brick base on which they placed a cement top with a doughnut shaped depression. They placed mercury in the depression and let a wooden float rest upon the mercury. On the wood was a stone base in which the parts of the interferometer were firmly fixed. All was so well balanced that the lightest touch would make the interferometer revolve steadily on its mercury support.

Now they were ready for what was to come to be known as the "Michelson-Morley experiment." Once again a ray of light was split and sent out on its errand; and once again the light failed and brought back nothing. The only interference fringes that were to be seen were tiny ones that clearly represented unavoidable imperfections of the instrument.

Of course, it might be that the rays of light weren't

heading exactly upwind and downwind, but in such directions that the ether wind had no effect. However, the instrument could be rotated. Michelson and Morley took measurements at all angles—surely the ether wind had to be blowing in one of those directions. They did even better than that. They kept taking measurements all year while the earth itself changed direction of motion constantly as it moved in its orbit about the sun.

They made thousands of observations, and by July 1887 they were ready to report. The results were negative. They had tried to measure earth's absolute velocity and they had failed and that was that.

There had to be an explanation of this failure and no less than five of them can be considered for a moment. I'll list them.

1) The experiment can be dismissed. Perhaps something was wrong in the equipment or the procedure or the reasoning behind it. Men such as the English scientists Lord Kelvin and Oliver Lodge took that point of view. However, this point of view is not tenable. Since 1887, numerous physicists have repeated the Michelson-Morley experiment with greater and greater precision. In 1960 masers (atomic clocks) were used for the purpose and an accuracy of one part in a trillion was achieved. But always, down to and including the 1960 experiments, the Michelson-Morley failure was repeated. There were no interference fringes. The light rays took precisely the same time to travel in any direction, regardless of the ether wind.

2) Well, then, the experiment is valid and shows there is no ether wind, for any one of four different reasons:

a) The earth is not moving; it is the motionless center of the universe. This would involve so many other paradoxes and would fly in the face of so much astronomical and physical knowledge gained since the time of Copernicus that no scientist seriously advanced this for a moment. However, a friend of mine has pointed out that the only way of disproving this suggestion beyond a doubt is to run the Michelson-Morley experiment elsewhere than on the earth. Perhaps when we reach the moon, we ought to make it one of the early orders of business to repeat the Michelson-Morley experiment there. If it proves negative (and I'm sure it will!), we can certainly conclude that the earth and the moon can't both simultaneously be motionless. That one or the other is motionless is, at least, conceivable; that both are, is not.

b) The earth does move, but in doing so it drags the neighboring ether with it, so that it seems motionless compared with the ether that is right at the surface of the earth. For that reason, no interference fringes are produced. The English physicist George Gabriel Stokes suggested this. Unfortunately, this implies that there is friction between the earth and the ether and this would raise the serious question as to why the motion of heavenly bodies wasn't continually being slowed by their passage through the ether. It was as hard to believe in the "ether drag" as in the motionless earth, and Stokes's notion died a quick death.

Two suggestions survived, however:

c) The Irish physicist George Francis Fitzgerald suggested that all objects (and therefore all measuring instruments) grew shorter in the direction of motion, according to a formula which was easily derived. This is the "Fitzgerald contraction." The Fitzgerald contraction introduced a factor that just neutralized the difference in time spent by the two light rays on their travel and therefore accounted for the absence of interference fringes. And yet the Fitzgerald contraction had the appearance of a "gimmick factor." It worked, yes, but why should the contraction exist at all?

d) The Austrian physicist Ernst Mach went to the heart of the matter. He said there were no interference fringes because there was no ether wind because there was no ether. What could be simpler.

This was not a strange thing for Mach to have said. He was a rebel who insisted that only observable phenomena were rightly a matter for scientific inquiry, and that scientists should not set up models that were not themselves directly observable, and then believe in their actual existence. Mach even refused to accept atoms as anything more than a convenient fiction. Naturally, it was to be expected that he would be ready to scrap the ether the first chance he got.

How tempting that must have been! The ether was such a ridiculous and self-contradictory substance that some of the greatest nineteenth-century theoretical physicists had worn themselves out trying to explain it. Why not throw it away, as Mach irascibly suggested?

The trouble was—how would one then account for the fact that light could cross a vacuum? Everyone admitted that light consisted of waves, and the waves had to be waves of *something*. If the ether existed, light consisted of waves of ether. If the ether did not exist, then light consisted of waves of *what?*

Physics was hovering between the frying pan of ether and the fire of complete chaos, and heaven only knows what would have happened if two German scientists, Max Karl Ernst Ludwig Planck in 1900, and Albert Einstein in 1905, had not come along to save the situation.

Save it, however, they did. The work of Planck and Einstein proved that light behaved as particles in some ways and that the ether therefore was not needed for light to travel through a vacuum. When this was done, the ether was no longer useful and it was dropped with a glad cry. The ether has never been required since. It does not exist now; in fact it never existed. (Einstein's work also placed the Fitzgerald contraction in the proper perspective.)

As a result, the Michelson-Morley light-that-failed was recognized as the most tremendously successful failure in the history of science, for it completely altered the physicists' view of the universe. In 1907 Michelson received the Nobel Prize in Physics, the first American to win one of the science prizes.

10. The Light Fantastic

WHEN I WAS YOUNG, we children used to listen to something called "radio." It's a hard thing to describe to the modern population, but if you imagine a television set with the picture tube permanently out of order, you've got the essentials.

On the radio set there was a dial you could turn in order

to tune in the various stations and the dial had markings numbered from 55 to 160. As far as I know, nobody I knew had any idea what those numbers meant—or cared.

A particular radio station might describe itself as possessing "880 kilocycles," and eventually I deduced that the numbers on the radio dial referred to tens of kilocycles, but again I never bumped into anyone (when I was young) who knew or cared what a kilocycle was.

In fact, as I look back upon it now, I don't think I knew or cared myself. I could dial any radio station I wanted with quick sureness and I had the radio schedule memorized. What more could I want?

And yet, if you consider the dial of a radio set, and proceed by free association, you can end up with some pretty amazing matters, as I shall try to show you.

I'll begin with waves.

The most important waves in the universe are set up by oscillating electric charges. Since all electric charges have associated magnetic fields, the radiating waves produced in this fashion are called "electromagnetic." Electromagnetic waves radiate outward from the point of origin, moving at the velocity of light—which is not surprising, since light is itself an electromagnetic radiation.

Each oscillation of the electric charge back and forth gives rise to a single wave and from this fact we can calculate the length of the wave to which it gives rise. The length of a wave is called, with commendable simplicity, the "wavelength," and it is usually symbolized by the Greek letter *lambda* (λ).

Now suppose the electric charge oscillates once per second. By the time the end of the wave is formed at the completion of the oscillation, the beginning of the wave has been speeding out through space at the velocity of light for one full second. The velocity of light in a vacuum is 186,200 miles per second or, in the metric system, which I shall use exclusively in this article, 300,000 kilometers per second. If, therefore, it takes a second to form the wave, the beginning of the wave is 300,000 kilometers ahead of the end of the wave and the wavelength is 300,000 kilometers.

Suppose the electric charge oscillates twice per second.

Then in one second two waves are formed. Together they stretch out over 300,000 kilometers and each wave is 150,000 kilometers long; 150,000 kilometers is therefore the wavelength.

If the electric charge oscillates ten times per second, each wave is 30,000 kilometers long. If it oscillates fifty times per second, the wavelength is 6000 kilometers, and so on.

The number of oscillations per second can be called the "frequency," and this is usually symbolized by the Greek letter *nu* (ν).

As you see, what I have been doing in order to work out the wavelength of electromagnetic radiation is to divide the velocity of light (usually represented by c) by the frequency of the radiation. Put this in equation form and you have:

$$\lambda = \frac{c}{\nu}$$

If you know the wavelength and want to find the frequency, you need only solve for ν in the equation above, and you have:

$$\nu = \frac{c}{\lambda}$$

Thus, if the wavelength is 15 kilometers, then the frequency is $\frac{300,000}{15}$, or 20,000 oscillations per second.

A frequency of one oscillation per second can be described as one cycle. A frequency of a thousand oscillations per second can be described as one kilocycle (the prefix *kilo-*, as I explained in Chapter 7, being used in the metric system to represent "one thousand"). If, then, radio station WNBC in New York is located at 660 kilocycles (or at 66 on the dial), then that means the wave it puts out has a frequency of 660,000 oscillations per second. The wavelength of those waves is $\frac{300,000}{660,000}$ or 0.455 kilometers. This is equivalent to 455 meters.

In the same way, we can calculate the wavelengths of the waves put out by some other New York radio stations:

115

	Kilocycles	*Wavelength*
		(*meters*)
WOR	710	425
WABC	770	390
WNYC	830	360
WCBS	880	340
WNEW	1130	265
WQXR	1560	190

Notice that the wavelength gets shorter as the kilocycles increase; which is why, if we go up high enough on the dial, we end up with "short-wave radio." One way of expressing this relationship is to say that frequency and wavelength are inversely proportional to each other; as one goes up, the other goes down.

An electromagnetic radiation can have any wavelength, as far as we know, since a charged particle can oscillate at any frequency. There is no upper limit to the wavelength, certainly, for the oscillation can be slowed down to zero, in which case the wavelength approaches the infinite.

On the other hand, electric charges can be made to oscillate millions of times per second by man. Atoms can (in effect) oscillate trillions of times a second. Electrons can oscillate quadrillions and even quintillions of times per second. Nuclear particles can oscillate sextillions and even septillions of times per second. Wavelengths can get shorter and shorter, with no lower limit in theory.

The properties of electromagnetic radiation vary with frequency. For one thing, the radiation is put out in discrete little bundles called "quanta" and the energy content of one quantum of a particular radiation is in direct proportion to its frequency. As frequency goes up (and wavelength down) the radiation becomes more energetic and can interact more thoroughly with matter.

Short-wave radiation may knock electrons out of metals where longer-wave, less energetic radiation will not, and this is known as the photoelectric effect. (Einstein explained the rationale behind the photoelectric effect in 1905, the same year in which he first advanced his theory of relativity; and when he got his Nobel Prize in 1921, it was for explaining the photoelectric effect, *not* for relativity.)

116

Again, short-wave radiation will bring about certain chemical changes where long-wave radiation will not, which is why you can develop ordinary photographic film under a red light. The red-light radiations are too low in energy to affect the negative.

Certain ranges of radiation are energetic enough to affect the retina of the eye and give us the sensation we call light. Radiation less energetic cannot be seen, but the energy can be absorbed by the skin and felt as heat. Radiation more energetic cannot be seen either, but can damage the retina and burn the skin.

It is convenient for physicists to divide the entire range of electromagnetic radiation (the "electromagnetic spectrum") into arbitrary regions. Here they are in the order of increasing frequency and energy and, therefore, of decreasing wavelength.

1) *Micropulsations.* These have frequencies of less than 1 cycle and, therefore, wavelengths of more than 300,000 kilometers. Such radiation has been detected with frequencies of as little as 0.01 cycle. This means that one oscillation takes 100 seconds and the wavelength is 30,000,000 kilometers, or three fourths of the way from here to Venus at its closest, which isn't bad for one wave.

2) *Radio waves.* In its broadest sense, these would include everything with frequencies from 1 cycle to 1 billion (10^9) cycles, and with wavelengths from 300,000 kilometers down to 30 centimeters. Actually, long-wave radio makes use of frequencies from 550,000 cycles to 1,600,000 cycles and wavelengths from 550 meters down to 185 meters. Short-wave radio uses wavelengths in the 30-meter range, and television in the 3-meter range.

3) *Microwaves.* The frequencies are from 1 billion (10^9) to 100 billion (10^{11}) cycles and the wavelengths are from 30 centimeters to 0.3 centimeters. The radiation detected by radio telescopes is in this range and the radiation of the neutral hydrogen atom (the famous "song of hydrogen") has a wavelength of 21 centimeters. Radar also makes use of this range.

4) *Infrared rays.* The frequencies are from 100 billion (10^{11}) cycles to nearly a quadrillion (10^{14} plus) cycles and the wavelengths run from 0.3 centimeters to 0.000076 centimeters. Infrared wavelengths are usually measured in

117

microns, one micron being a ten-thousandth of a centimeter, so the infrared wavelength range can be said to extend from 3000 microns down to 0.76 microns.

5) *Visible light rays.* These include a short stretch of frequencies just under the quadrillion mark (10^{15} minus), with wavelengths from 0.76 microns to 0.38 microns. Light wavelengths are usually measured in angstrom units, one angstrom unit being equal to a ten-thousandth of a micron. Thus, the wavelengths of visible light range from 7600 angstrom units down to 3800 angstrom units.

6) *Ultraviolet rays.* These include frequencies from a quadrillion (10^{15}) cycles up to nearly a hundred quadrillion (10^{17} minus) cycles, and the wavelengths run from 3800 angstrom units down to about 100 angstrom units.

7) *X-rays.* These include frequencies from nearly a hundred quadrillion (10^{17} minus) up to a hundred quintillion (10^{20}) cycles, wih wavelengths ranging from 100 angstrom units down to 0.1 angstrom units.

8) *Gamma rays.* These make up the frequencies that are more than a hundred quintillion (10^{20}) cycles and wavelengths less than 0.1 angstrom units.

Actually, the dividing lines are anything but sharp, and X-rays and gamma rays, in particular, overlap generously. People speak of a particular frequency as being an X-ray if it is created in an X-ray tube and as a gamma ray if it is produced by a nuclear reaction. You can have soft gamma rays with wavelengths some three hundred times as long as the hardest X-rays. However, a particular wavelength has a particular energy and a particular set of properties regardless of what you call it: X-ray, gamma ray, or herring. By setting the boundary between X-rays and gamma rays at a frequency of a hundred quintillion cycles, I merely cut the overlap in half and am perfectly willing to admit the boundary is arbitrary.

Now this is a bewildering array of frequencies and wavelengths and I wouldn't be me if I didn't look for an easier way of presenting it. The easier way is drawn from usage in connection with sound waves. Sound waves are not electromagnetic in nature, but they, too, have wavelengths and frequencies.

We detect differences in the frequency of sound waves, at least in the audible range, by differences in the pitch we hear. It is conventional in our culture to write music

118

using a series of notes with fixed frequencies. I will begin with the note on the piano which is called "middle C" and give its frequency and that of successive notes as we proceed toward the right on the keyboard, considering white keys only:

do — 264	do — 528	do — 1056
re — 297	re — 594	
mi — 330	mi — 660	
fa — 352	fa — 704	
sol — 396	sol — 792	
la — 440	la — 880	
ti — 495	ti — 990	

Notice that the frequency of each "do" is just twice the frequency of the preceding one. In fact, starting anywhere on the keyboard, one can progress through seven notes of increasing frequency and end with an eighth note of just twice the frequency of the first. Such a stretch is called an "octave," from the Latin word for "eight."

Applying this to any wave form in general, one can speak of an octave as applying to any continuous region stretching from a frequency of x to one of $2x$. Since wavelength is inversely proportional to frequency, every time a frequency is doubled, a wavelength is halved. Every region stretching from a wavelength of y to one of $y/2$ is also an octave, therefore.

So we can break up the electromagnetic spectrum into octaves. As an example, the longest wavelengths of visible light are 7600 angstrom units, while the shortest are 3800 angstrom units. The shortest wavelengths are just half the longest and so the range covered by visible light is equal to one octave of the electromagnetic spectrum.

Since there is no upper or lower limit to the frequencies of the electromagnetic spectrum, the number of octaves is, theoretically, infinite. However, suppose we consider a wavelength of 30,000,000 kilometers as the practical maximum, since this is the longest micropulsation detected, and a wavelength of 0.0001 angstrom units as the practical minimum, since beyond that lie the energy ranges associated with cosmic rays, which are particulate rather than electromagnetic in nature.

The number of times you must halve 30,000,000 kilometers to reach 0.0001 angstrom units is 81. (Try it and see, and remember that 1 kilometer equals 10,000,000,-000,000 angstrom units.) The portion of the electromagnetic spectrum I have marked off, therefore, is eighty-one octaves long, and of that length, we see exactly one octave with our eyes.

Now let's measure off the confusing divisions of the electromagnetic spectrum in octaves and the picture will be much simpler:

	Octaves
micropulsations	6½
radio waves	30
microwaves	6½
infrared rays	12
visible light rays	1
ultraviolet rays	5
X-rays	10
gamma rays	10
total	81

As you see, two thirds of the octaves are longer-wave and, therefore, less energetic than light. In fact, the radio-wave region at its broadest takes up a third of the octaves of the spectrum. Actually, though, only about twelve octaves altogether are used for radio and television communications.

Still that makes up about 15 percent of the total number of octaves and as our needs for communication increase with the developing space age, how much room for expansion can there be?

The answer is: Plenty!

To see why that is, let's consider this matter of octaves further. In the realm of sound, the ear finds all octaves equal. In each one, there is room for seven different notes (plus sharps and flats, of course) before the next octave begins.

This is not so, however, as far as communication by electromagnetic waves is concerned. As one goes up the electromagnetic spectrum in the direction of increasing frequency, each octave has more room than the one before.

Each television channel emits a carrier wave which it modifies, these modifications being converted into sight and sound at the receiving television set. In order for two channels not to interfere with each other, they must have frequencies that are not too close. They can't be anywhere near as closely spaced as the radio stations with which I began this article, for instance. The width of a standard television channel is 4,000,000 cycles (or 4 megacycles, a megacycle being equal to a million cycles).

The television channels fall at the short-wave end of the radio-wave region, in the range of frequencies of 100,000,000 cycles (100 megacycles) and wavelengths of about 3 meters.

Consider an octave in this region of frequencies; say a stretch of the spectrum from a frequency of 80 megacycles to one of 160 megacycles. This covers a width of 80 megacycles, and if television channels are spaced at 4 megacycle intervals, there is room for twenty channels.

In the next octave up, from 160 to 320 megacycles, there is room for forty channels. In the one after that, from 320 to 640 megacycles, there is room for eighty channels.

The number of television channels per octave of electromagnetic radiation doubles for each successive octave as one moves up the scale in the direction of increasing frequency. In fact, each octave of electromagnetic radiation contains about as much room for television channels as do all the preceding octaves put together.

What about visible light, then? There is only one octave of visible light, but it is roughly twenty-two octaves higher than the one used for television. There is thus 2^{22} times as much room for television channels in the octave of light as in the octave ordinarily used for television. The figure 2^{22} represents the product of twenty-two 2's, and that comes to over four million. (Multiply them out for yourself; don't take my word for it.)

In other words, for every channel available in the usual television portion of the electromagnetic spectrum, there would be some four million channels available in the visible light portion.

We can break this down in more detail. The visible spectrum contains a number of colors that fade one into the other as you go up or down the scale. Actually, the eye can distinguish among a great number of shades and there are no sharp boundaries. Nevertheless, it is customary to divide the visible spectrum into six colors, which, in order of increasing frequency, are red, orange, yellow, green, blue, and violet. And each color is considered as stretching over a certain range of frequency. The situation might be presented thus:

	Wavelength Range (angstrom units)	Frequency Range (megacycles)
red	7600 to 6300	400,000,000 to 475,000,000
orange	6300 to 5900	475,000,000 to 510,000,000
yellow	5900 to 5600	510,000,000 to 540,000,000
green	5600 to 4900	540,000,000 to 615,000,000
blue	4900 to 4500	615,000,000 to 670,000,000
violet	4500 to 3800	670,000,000 to 800,000,000

Remembering that the width of a standard television channel is only 4 megacycles, then we can set up the following table:

	Width of Frequency Band (megacycles)	Number of Television Channels Possible
red	75,000,000	19,000,000
orange	35,000,000	9,000,000
yellow	30,000,000	7,000,000
green	75,000,000	19,000,000
blue	55,000,000	14,000,000
violet	130,000,000	32,000,000
	Total	100,000,000

Well, then, why not use light waves as carriers for television broadcasts?

Until two years ago this was a suggestion that could have only theoretical interest. The carrier waves set up for ordinary radio-television communication can be produced in perfect phase. They form an orderly succession of waves that can be neatly modified in any fashion.

Light waves, however, cannot be set up so neatly in phase; at least they couldn't until the 1960's. It is quite impractical to try to oscillate an electric circuit five hundred trillion times a second, which is what would be required to send out a beam of visible light. The electrons within an atom must be relied upon for such an oscillation. Heat is poured into them and it is liberated as electromagnetic radiation, much of which (because of their natural rate of oscillation) is in the form of visible light. In other words, you can make light by starting a fire.

The only trouble is that the various heated atoms give off radiation, each in its own good time, and the wavelength is not fixed but can be varied over a wide range, and the quantum is fired out in any direction. Thus, the emitted light waves are so much out of phase that most of their energy is canceled and converted into heat; they spread out widely in every direction and cover a broad range of the spectrum. In short, the light produced is good enough to see by, but not good enough to serve as a carrier wave for TV.

However, in 1960 instruments were devised into which energy could be pumped and then, when a sparking bit of light was allowed to enter, all the energy was converted into light of the same wavelength, and all in phase. The device could be so constructed that all the light would emerge in the same direction, too.

The beam of intense light that is produced by such a device would stick together (it would be "coherent") and it would possess an extremely narrow band of wavelengths (it would be "monochromatic"). The process by which a bit of light sparks the conversion of energy into a lot of light is called "*l*ight *a*mplification by *s*timulated *e*mission of *r*adiation," and by taking the indicated initials, the instrument was named a "laser." (In case you are interested, a word constructed out of the initials of a phrase is called an "acronym.")

Of course, even so, the use of light as television carrier waves presents difficulties. In the range of the electromagnetic spectrum currently used for television, the radiation can penetrate buildings and go through ordinary obstacles. Visible light can't do this. You would need a clear and unobstructed view of the TV station before you could receive a program.

It is possible, however, that light might be sent through plastic pipes, from which leads could reach each television

123

set in the area. (Does that mean the streets all get dug up, or will the pipes run along telephone poles, or what?)

On the other hand, television by laser would be ideal out in space, where ship could reach ship or space station through the uninterrupted reaches of vacuum, and each ship could have a television channel reserved all for itself. It would be a long time before we had more than a hundred million ships out in space, so there would be no crowding. Then, even if we did run out of room in the visible region, the ultraviolet portion of the spectrum would give room for about six billion more channels.

Of course, there is something else—

These days, when I watch television here at home, I have my choice of four channels that I can get with reasonable clearness and audibility. Even with only four channels at their disposal, however, the television moguls can supply me with a tremendous quantity of rubbish.

Imagine what the keen minds of our entertainment industry could do if they realized they had a hundred million channels into which they could funnel new and undreamed-of varieties of trash.

Maybe we ought to stop right now!

Part Three

CHEMISTRY

11. Slow Burn

FOR MANY YEARS NOW I have been an inveterate admirer of Sir Isaac Newton. One can, after all, make out a good case for his having been the greatest scientist who ever lived.

What's more, it doesn't displease me one little bit that Newton's first name is Isaac. To be sure, I wasn't named for him, but for my grandfather. Yet the principle remains; we have something in common. And to top it off, the Boston suburb in which I live is named Newton—how do you like that?

So you see, I have lots of reasons for being an Isaac Newton fan and it therefore pains me to admit there are flaws in the shining picture he presents. In physics and astronomy he was a transcendent genius. In mathematics he was a ground-breaking prodigy. Yet in chemistry he was nothing but a bumbler. He wasted his time in a vain and useless effort to manufacture gold, scouring Europe for recipes, trying each one and forever being disappointed.

This is a dramatic way of showing that Newton stood at a midway point in the history of the physical sciences. In the 1680s when he announced his laws of motion and his theory of gravitation, the birth of modern physics (thanks to Galileo) was just one century in the past and the birth of modern chemistry (thanks to Lavoisier) was just one century in the future.

Now the story of the birth of physics has been told

and told again. We all know (or should) about Galileo's experiments with falling bodies which, at one stroke, destroyed Aristotelian physics and established the modern form of the science. In popular mythology this is concentrated into a single experiment, the dropping of a heavy and light ball from the top of the Leaning Tower of Pisa and watching them hit the ground simultaneously. (Actually, it is quite certain that Galileo never performed this experiment.)

On the other hand, the birth of chemistry is graced by no such key experiment. There is no chemical equivalent of dropping weights off the Leaning Tower of Pisa; no single, classic feat to go ringing down the corridors of time as the smasher of the old and the beginner of the new. At least, I don't find one in the books I've read on the subject; not one that is pointed to as *the* experiment.

Except that I think I've found one. I think I can make a case for the existence of a single, simple experiment that smashed the old chemistry and started the new chemistry. It was every bit as dramatic and conclusive (if not quite as spectacular) as the Leaning Tower of Pisa experiment, except that:

1) The crucial chemical experiment really happened and is not a myth, and

2) It involved a mad scientist and should therefore strike a nostalgic chord in the hearts of all true science-fiction fans.

With your permission then, O Gentle Reader (or, if necessary, without), I shall tell the story of the birth of Modern Chemistry, as I see it.

In the time of Newton chemical theory was still based, in large part, on what the Greek philosophers had worked out two thousand years earlier. The "four elements" (that is, the fundamental susbtances out of which the universe was made) were earth, water, air, and fire.

The Greek philosophers felt that actual bodies were made up of the four elements in particular proportions. One could well imagine, then, that the elements in one body could be separated and then recombined in different proportions to form a second body of a different sort. In this way, one could change one metal into another (if one

126

could but discover the correct procedure), and in particular, one could change lead into gold.

For about fifteen hundred years, alchemists tried to find out the proper recipe for such "transmutation." The Arabs, in the process, worked out the theory that there were two special principles involved in the different solid bodies with which they worked. There was the metallic principle, mercury, and the combustible principle, sulfur.

This didn't help them make gold, and by Newton's time chemistry seemed badly in need of some new ideas. What's more, any new ideas that did come along ought to deal with combustion. Coal was beginning to come into use as a new fuel. Men were beginning to play with the steam produced by the heat of burning fuel. In general, the matter of combustion was in the air and as exciting in 1700 as electricity was to be in 1800, radioactivity in 1900, and rocketry in 1950.

Onto the scene then, steps a German physician named Georg Ernest Stahl. While still in his twenties he was appointed court physician to the Duke of Weimar. In later life he was to become physician to still higher royalty, King Frederick William I of Prussia. His lectures on medicine at the University of Halle were famous and well attended.

In 1700 this man advanced a theory of combustion that made more sense than anything previously suggested. He drew heavily on alchemical notions and, in particular, on the combustible principle, sulfur. He gave this principle a new name and described its behavior in greater detail.

The principle he called "phlogiston," from a Greek word meaning "to set on fire," for he held that all inflammable objects contained phlogiston and it was only the phlogiston content that made it possible for them to burn.

During the process of combustion, said Stahl, the burning material lost its content of phlogiston, which poured out into and was received by the air. What was left after combustion was completely lacking in phlogiston and could burn no more. Wood and coal, for instance, were rich in phlogiston, but the ash they left behind contained none.

Stahl's greatest contribution to chemical thinking was his suggestion that the process of rusting of metals was similar in principle to that of the burning of wood. A metal, such as iron, was rich in phlogiston. As it corroded, it lost phlogiston to the air, and when all the phlogiston was gone, only rust was left behind.

The basic difference, then, between the burning of wood and the rusting of iron was no more than a matter of speed. Wood lost phlogiston so rapidly that the velocity of its passing made it visible as flame. Iron lost phlogiston so slowly that its passage was imperceptible. Burning, in Stahl's view, was a fast rusting, while rusting was a slow burn.

In this, Stahl was quite correct, but he gets little credit for it. About the first thing chemistry students are taught to do is to laugh at the phlogiston theory, so that Stahl is either forgotten or condemned, and I consider that unfair.

As a matter of fact, the phlogiston theory explained quite a few things that were not explained before, most notably the matter of metallurgy. For instance, it had been known for thousands of years that if metal ore were heated strongly, in contact with burning wood or charcoal, the free metal could be obtained. As for *why* this happened, no one had a good answer.

Until Stahl, that is. According to the phlogiston theory, it was easy to see that a metal ore was a kind of naturally occurring rust that was completely free of phlogiston and therefore showed no metallic properties. If heated in the presence of phlogiston-rich charcoal, phlogiston passed from the charcoal to the ore. As the ore gained phlogiston, it turned into metal. As the charcoal lost phlogiston, it turned into ash.

Isn't that neat?

Unfortunately, there was one great flaw in the theory. When a metal rusted, it gained weight! One pound of iron produced about one and a half pounds of iron rust. If the conversion were the result of the loss of phlogiston and not the gaining of anything, where did the extra weight come from?

A few chemists worried about this and tried to explain that phlogiston had negative weight! Instead of phlogiston being pulled down by gravity, it was pushed up by levity. (You may take that as a pun, if you choose, but levity was the actual term used.) Thus, a pound of iron could be considered as containing minus half a pound of phlogiston, and when the phlogiston left, the resulting rust would weigh one and a half pounds.

This notion went over like a lead balloon. For one thing, no example of levity was found anywhere in nature outside the phlogiston, and for another, when wood burned it *lost* weight. The ash it left behind was much lighter than the original wood. If the wood had lost phlogiston and if phlogiston exerted a force of levity, why wasn't the ash heavier than the wood, as rust is heavier than iron?

There was no answer to this, and the average chemist of the day simply shrugged. There was, after all, no tradition of exact measurement in chemistry. For thousands of years everyone had worked the chemical industries as art forms rather than as sciences. The alchemists had involved themselves in purely descriptive observations. They had noted the formation of precipitates, the emission of vapors, the changes of colors—but such things as weight and volume were irrelevant.

For two generations matters continued thus, and then, in the 1770s, a number of momentous developments took place. For one thing, chemists began to concern themselves with air.

To the ancient Greeks air was an element, a single substance. However, the Scottish chemist Joseph Black burned a candle in a closed container of air, as the 1770s opened, and found that the candle eventually went out. When it did, there was still plenty of air in the container, so *why* did it go out?

He was busy with other matters, so he passed the problem on to a student of his named Daniel Rutherford. (Rutherford, by the way, was the uncle of the poet and novelist Sir Walter Scott.)

In 1772 Rutherford repeated Black's experiments and went further. New candles, set on fire and placed in the air remaining after the old candle had burned out, promptly went out themselves. Mice, placed in such air, died.

Rutherford analyzed these observations in terms of the phlogiston theory. When a candle burned in an enclosed volume of air, it gave up phlogiston to the air but, apparently, any given volume of air could only hold so much phlogiston and no more. When the air was filled with phlogiston, the candle went out and nothing further would burn in that air. A living creature which, in the process of breathing, constantly gave up phlogiston (there had been speculations dating back to Roman times that respiration was analogous to combustion) could not do so in this phlogiston-filled air, and died. Rutherford called this asphyxiating gas "phlogisticated air."

The scene now shifts southward to England, where a Unitarian minister Joseph Priestley had become interested in science after he met the American scientist and statesman Benjamin Franklin in 1766.

Priestley's great discovery came from experiments with mercury in 1774. He began by heating mercury with sunlight concentrated through a large magnifying glass. The heat caused the gleaming surface of the mercury to be coated with a reddish powder. Priestley skimmed off the powder and heated it to a still higher temperature. The powder evaporated, forming two different gases. One of these was mercury vapor, for it condensed into droplets of mercury in the cool upper regions of the vessel. The other remained an invisible vapor.

How did Priestley know it was there? Well, it had peculiar properties that were not like those of ordinary air. A smoldering splint of wood thrust into the container in which the red powder from mercury was being heated burst into bright flame. Priestley collected the vapors and found a candle would burn in it with unearthly brightness; he found that mice placed in the vapor would jump about actively; he even breathed some himself and reported it made him feel very "light and easy."

Priestley interpreted all this according to the phlogiston theory. When mercury was heated it lost some of its phlogiston to air and became a red powder which lacked phlogiston and could be considered a kind of mercury rust. If he heated this mercury rust strongly, it absorbed phlogiston from the air and became mercury again. Meanwhile, the air in the neighborhood was bled of its phlogiston and became "dephlogisticated air." Naturally, such dephlogisticated air was unusually thirsty for phlogiston. It sucked phlogiston rapidly out of a smoldering splint and the velocity of the reaction was visible as a burst of flame. For similar reasons, candles burned more brightly and mice ran about more actively in dephlogisticated air than in ordinary air.

The Priestley and Rutherford experiments, taken together, seemed to show that air was a single material substance, which could be altered in properties by a variation in its content of the imponderable fluid, phlogiston.

Ordinary air contains some phlogiston but is not saturated with it. It can gain phlogiston when something burns in it; or it can lose phlogiston when a rust heated in it becomes a metal. When it gains all the phlogiston it can hold, it will no longer support combustion or life and it is then Rutherford's gas. If it loses all the phlogiston it has, it will support combustion with great eagerness and life with great ease and will then be Priestley's gas.

Now we shift still farther south. In Paris the brilliant young chemist Lavoisier is working hard under the stress of an idea—that measurement is as important to chemistry as Galileo showed it to be to physics. Qualitative observations are insufficient; one must be quantitative.

As an example—when water, even the purest, was slowly boiled away in a glass vessel, some sediment was always left behind. Alchemists had often done this and they had pointed to the sediment as an example of the manner in which the element water had been converted to the element earth. (From this they deduced that transmutation was possible and that lead could be turned to gold.)

About 1770 Lavoisier decided to repeat the experiment, but quantitatively. He began by accurately weighing a clean flask and adding an accurate weight of water. He then boiled the water under conditions so designed that the rising water vapor was cooled, condensed back to water, and forced to drip again into the still-boiling contents of the flask. He continued this for 101 days, thus giving the water plenty of time to turn into earth. He then stopped and let all the water cool down.

Sure enough, as the water cooled, the sediment formed. Lavoisier poured out the water, filtered off the sediment, and weighed each separately. The weight of the water had not changed at all. He then weighed the flask. The flask had lost weight and the loss in weight was just equal to the weight of the sediment. Water had not changed to earth; it had simply dissolved some of the material of the flask.

Thus he showed that one conclusion drawn from a particular experiment could be shifted to another and much more plausible conclusion by simply becoming quantitative.

In a later experiment Lavoisier put some tin in a vessel which he then closed. He next weighed the whole business accurately. Then he heated the vessel.

A white rust formed on the tin. It was known that such a rust was invariably heavier than the original metal, yet when Lavoisier weighed the whole setup, he found the total weight had not changed at all. If the rust were heavier than the tin, then that gain in weight must have been countered by an equal loss in weight elsewhere in the vessel. If the loss in weight were in the air content, then a partial vacuum should now exist in the vessel. Sure enough, when Lavoisier opened the vessel, air rushed in and then the system increased in weight. The increase was equal to the extra weight of the rust.

Lavoisier therefore suggested the following: Combustion (or rust-formation) was caused not by the loss of phlogiston but by the combination of the fuel or metal with air. Phlogiston had nothing to do with it. Phlogiston did not exist.

The weak point in this new suggestion, just at first, lay in the fact that not all the air was involved in this. Lavoisier found that when a candle burned, it used up only about one fifth of the air. It would burn no longer in the remaining four fifths.

Light dawned when Priestley visited France and had a conversation with Lavoisier. Of course! Lavoisier rushed back to his work. If phlogiston did not exist, then air could not change its properties with gain or loss of phlogiston. If two kinds of air seemed to exist with different properties, then it was because air contained two different substances.

The one fifth of the air which a burning candle used up was Priestley's dephlogisticated air, which Lavoisier now called "oxygen," from Greek words meaning "sourness-producer." (Lavoisier thought oxygen was a necessary component of acids. It isn't, but the name will never be changed now.) As for the remaining four fifths of the air, that portion in which candles would not burn and mice would not live, that was Rutherford's phlogisticated air, and Lavoisier called it "azote," from Greek words meaning "no life." Nowadays, we call it "nitrogen."

Air, according to Lavoisier, then, was one fifth oxygen and four fifths nitrogen. Combustion and rusting were brought about by the combination of materials with oxygen only. Some combinations (or "oxides"), such as carbon dioxide, were vapors and left the scene of combustion altogether, which was why coal, wood, and candles all lost weight drastically after burning. Other oxides were solids and remained on the spot, which was why rust was heavier than metal—heavier by the added oxygen.

In order for a new theory to displace an old comfortable one, the new theory has to be *clearly* better, and the oxygen theory was not, just at first. To most chemists, oxygen just seemed phlogiston in reverse. Instead of wood losing phlogiston in combustion, it gained oxygen. Instead of iron ore gaining phlogiston in iron smelting, it lost oxygen.

Lavoisier could only have carried conviction by proving that the matter of weight was crucial, for the oxygen theory explained the weight changes in combustion and rusting, while the phlogiston theory did not and could not.

Lavoisier tried to emphasize the importance of weight and to make it central to chemistry by maintaining that there was no change in total weight during the course of any chemical reaction in a *closed* system, where vapors were not allowed to escape and outside air could not be added. This is the "law of conservation of mass." Another way of putting it is that matter can neither be created nor destroyed, and if that is true, then the phlogiston theory is fallacious, for in it the added weight of the rust appears out of nowhere and matter must therefore be created.

Unfortunately, Lavoisier could not make the law of conservation of mass hard and fast at first. There was a flaw. Lavoisier tried to measure the amount of oxygen a human being absorbed in breathing and to compare it with the carbon dioxide he exhaled. When he did that, it always turned out that some of the oxygen had disappeared. The exhaled carbon dioxide never accounted for all the oxygen taken in. If the law of conservation of mass didn't hold, there was no handy stick with which to kill phlogiston.

Now let's go back to England and to our mad scientist, Henry Cavendish.

Cavendish, you see, was pathologically shy and unbelievably absentminded. It was only with difficulty that he could speak to one man; and it was virtually impossible to speak to more than one. Although he regularly attended dinner at the Royal Society, dressed in snuffy, old-fashioned clothes, he ate in dead silence with his eyes on his plate.

He was a woman-hater (or, perhaps, woman-fearer) to the point where he could not bear to look at one. He communicated with his female servants by notes, and any who accidentally crossed his path in his house was fired on the spot. He built a separate entrance to his house so he could come and leave alone. In the end, he even insisted on dying alone.

He came of a noble family and at the age of forty inherited a fortune, but paid no particular attention to it. Money literally meant nothing to him, and neither did fame. Many of his important discoveries he never bothered publishing, and it is only by going through the papers he left behind that we know of them.

Some discoveries, however, he did publish. In 1766, for instance, he discovered an inflammable gas produced by the action of acids on metals. This had been done before, but Cavendish was the first to study the gas systematically and so he gets credit for its discovery.

One thing that Cavendish noted about the gas was that it was exceedingly light—far lighter than air; lighter than any material object then known (or since discovered). With his mind on the "levity" that some had suggested as one of the properties of phlogiston, Cavendish began to wonder whether he had stumbled on something that was mostly, or even entirely, phlogiston. Perhaps he had phlogiston itself.

After all, as the gas left the metal through the action of acids, the metal formed a rust with phenomenal rapidity. Furthermore, the gas was highly inflammable; indeed, explosively so; and surely that was to be expected of phlogiston.

When, in the decade that followed, Rutherford isolated his phlogisticated air and Priestley his dephlogisticated air, it occurred to Cavendish that he could perform a crucial experiment.

He could add his phlogiston to a sample of dephlogisticated air and convert it first into ordinary air and then into phlogisticated air. If he did that, it would be ample proof that his inflammable gas was indeed phlogiston and, moreover, it would be a general proof of the truth of the phlogiston theory.

So, in 1781, Cavendish performed *the* crucial experiment in chemistry. It was simplicity itself. He merely set acid to working on metal so that a jet of his phlogiston could be forced out of a glass tube. This jet of phlogiston could be lighted by a spark and allowed to burn inside a vessel full of dephlogisticated air. That was all there was to it.

But when he did it, he found to his surprise that he had not formed phlogisticated air at all. Instead, the inner walls of the vessel were bedewed with drops of a liquid that looked like water, tasted like water, felt like water, had all the chemical properties of water and, egad, sir, *was* water.

Cavendish hadn't proved the phlogiston theory at all. In fact, as Lavoisier saw at once, Cavendish's experiment had once and for all killed phlogiston.

As soon as Lavoisier heard of Cavendish's work, he jumped upon it with loud cries of delight. He repeated the experiment with improvements and named Cavendish's gas "hydrogen," from Greek words meaning "water-producer," a name it keeps to this day.

Here's what this one simple experiment of Cavendish's did:

1) It proved water to be an oxide; the oxide of hydrogen. This was the last, final blow to the "four-elements" theory of the Greeks, for water was not a basic substance after all.

2) It wiped out the notion that air was a single substance varying in properties according to its phlogiston content. If that were so, then hydrogen plus oxygen would yield nitrogen (as Cavendish had, in truth, expected it would—using the eighteenth-century terminology of phlogisticated air, dephlogisticated air, and so on). But if air were not a single substance, then the only way of accounting for the experiments of the 1770s was to assume it a mixture of two substances.

3) Lavoisier realized that the foodstuffs that underwent combustion in the body contained both carbon *and* hydrogen. In the light of Cavendish's experiment, then, it was not surprising that the carbon dioxide produced by the body was less than sufficient to account for the oxygen absorbed. Some of the oxygen was used up in combining with hydrogen to form water, and expired breath was rich in water as well as in carbon dioxide. The obvious flaw in the law of conservation of mass was removed. The importance of quantitative measurement in chemistry was thus established and has never since been doubted.

In short, then, all of Modern Chemistry traces back, clean and true as an arrow, to Cavendish's burning jet of hydrogen.

There is an ironic postscript to the story, though. Lavoisier had one flaw in an otherwise admirable character. He had a tendency to grab for credit that did not belong to him. In advancing his theory of combustion, for instance, he never mentioned Priestley's experiments and never indicated that he had discussed them with Priestley. In fact, he tried to give the impression that he, himself, was the discoverer of oxygen. In the same way, when he repeated Cavendish's experiment of burning hydrogen, he tried to give the impression, without quite saying so, that the experiment was original with him.

Lavoisier didn't get away with these little tricks and posterity has forgiven him his vanity, for what he *did* do (including a deal of material I haven't mentioned in this article) was quite enough for a hundred ordinary chemists.

However, it is quite likely that neither Priestley nor Cavendish felt particularly kindly toward Lavoisier as a result. At least, neither man would accept Lavoisier's new chemistry. Both men refused to abandon phlogiston, and remained stubborn devotees of the old chemistry to the end of their lives.

Which once again proves, I suppose, that scientists are human. Like the metals they work with, they can be subjected to the effects of a slow burn.

12. You, Too, Can Speak Gaelic

IT IS DIFFICULT to prove to the man in the street that one is a chemist. At least, when one is a chemist after my fashion (strictly armchair).

Faced with a miscellaneous stain on a garment of unknown composition, I am helpless. I say "Have you tried a dry cleaner?" with a rising inflection that disillusions everyone within earshot at once. I cannot look at a paste of dubious composition and tell what it is good for just by smelling it; and I haven't the foggiest notion what a drug, identified only by trade name, may have in it.

It is not long, in short, before the eyebrows move upward, the wise smiles shoot from lip to lip, and the hoarse whispers begin: "Some chemist! Wonder what barber college *he* went to?"

There is nothing to do but wait. Sooner or later, on some breakfast-cereal box, on some pill dispenser, on some bottle of lotion, there will appear an eighteen-syllable name of a chemical. Then, making sure I have a moment of silence, I will say carelessly, "Ah, yes," and rattle it off like a machine gun, reducing everyone for miles around to stunned amazement.

Because, you see, no matter how inept I may be at the practical aspects of chemistry, I speak the language fluently.

But, alas, I have a confession to make. It isn't hard to speak chemistry. It just looks hard because organic chemistry (that branch of chemistry with the richest supply of nutcracker names) was virtually a German monopoly in the nineteenth century. The Germans, for some reason known only to themselves, push words together and eradicate all traces of any seam between them. What we would express as a phrase, they treat as one interminable word. They did this to the names of their organic compounds

137

and in English those names were slavishly adopted with minimum change.

It is for that reason, then, that you can come up to a perfectly respectable compound which, to all appearances, is just lying there, harming no one, and find that it has a name like para-dimethylaminobenzaldehyde. (And that is rather short, as such names go.)

To the average person, used to words of a respectable size, this conglomeration of letters is offensive and irritating, but actually, if you tackle it from the front and work your way slowly toward the back, it isn't bad. Pronounce it this way: PA-ruh-dy-METH-il-a-MEE-noh-ben-ZAL-duh-hide. If you accent the capitalized syllables, you will discover that after a while you can say it rapidly and without trouble and can impress your friends no end.

What's more, now that you can say the word, you will appreciate something that once happened to me. I was introduced to this particular compound some years ago, because when dissolved in hydrochloric acid, it is used to test for the presence of a compound called glucosamine and this was something I earnestly yearned to do at the time.

So I went to the reagent shelf and said to someone, "Do we have any para-dimethylaminobenzaldehyde?"

And he said, "What you mean is PA-ruh-dy-METH-il-a-MEE-noh-ben-ZAL-duh-hide," and he sang it to the tune of the "Irish Washerwoman."

If you don't know the tune of the "Irish Washerwoman," all I can say is that it is an Irish jig; in fact, it is *the* Irish jig; if you heard it, you would know it. I venture to say that if you know only one Irish jig, or if you try to make up an Irish jig, that's the one.

It goes: DUM-dee-dee-DUM-dee-dee-DUM-dee-dee-DUM-dee-dee, and so on almost indefinitely.

For a moment I was flabbergasted and then, realizing the enormity of having someone dare be whimsical at my expense, I said, "Of course! It's dactylic tetrameter."

"What?" he said.

I explained. A dactyl is a set of three syllables of which the first is accented and the next two are not, and a line of verse is dactylic tetrameter when four such sets of syllables occur in it. Anything in dactylic feet can be sung to the tune of the "Irish Washerwoman." You can sing most

138

of Longfellow's "Evangeline" to it, for instance, and I promptly gave the fellow a sample:

"THIS is the FO-rest pri-ME-val. The MUR-muring PINES and the HEM-locks—" and so on and so on.

He was walking away from me by then, but I followed him at a half run. In fact, I went on, anything in iambic feet can be sung to the tune of Dvorak's "Humoresque." You know the one—dee-DUM-dee-DUM-dee-DUM-dee-DUM-dee-DUM—and so on forever.)

For instance, I said, you could sing Portia's speech to the "Humoresque" like this: "The QUALiTY of MERcy IS not STRAINED it DROPpeth AS the GENtle RAIN from HEAV'N uPON the PLACE beNEATH."

He got away from me by then and didn't show up at work again for days, and served him right.

However, I didn't get off scot free myself. Don't think it. I was haunted for weeks by those drumming dactylic feet. PA-ruh-dy-METH-il-a-MEE-noh-ben-ZAL-duh-hide-PA-ruh-dy-METH-il-a-MEE-noh—went my brain over and over. It scrambled my thoughts, interfered with my sleep, and reduced me to mumbling semimadness, for I would go about muttering it savagely under my breath to the alarm of all innocent bystanders.

Finally, the whole thing was exorcized and it came about in this fashion. I was standing at the desk of a receptionist waiting for a chance to give her my name in order that I might get in to see somebody. She was a very pretty Irish receptionist and so I was in no hurry because the individual I was to see was very masculine and I preferred the receptionist. So I waited patiently and smiled at her; and then her patent Irish stirred that drumbeat memory in my mind, so that I sang in a soft voice (without even realizing what I was doing) PA-ruh-dy-METH-il-a-MEE-noh-ben-ZAL-duh-hide . . . through several rapid choruses.

And the receptionist clapped her hands together in delight and cried out, *"Oh, my, you know it in the original Gaelic!"*

What could I do? I smiled modestly and had her announce me as Isaac O'Asimov.

From that day to this I haven't sung it once except in telling this story. It was gone, for after all, folks, in my heart I know I don't know one word of Gaelic.

But what are these syllables that sound so Gaelic? Let's

trace them to their lair, one by one, and make sense of them, if we can. Perhaps you will then find that you, too, can speak Gaelic.

Let's begin with a tree of southeast Asia, one that is chiefly found in Sumatra and Java. It exudes a reddish-brown resin that, on being burned, yields a pleasant odor. Arab traders had penetrated the Indian Ocean and its various shores during medieval times and had brought back this resin, which they called "Javanese incense." Of course, they called it that in Arabic, so that the phrase came out *"luban javi."*

When the Europeans picked up the substance from Arabic traders, the Arabic name was just a collection of nonsense syllables to them. The first syllable *lu* sounded as though it might be the definite article (*lo* is one of the words for *the* in Italian; *le* and *la* are *the* in French and so on). Consequently, the European traders thought of the substance as "the banjavi" or simply as "banjavi."

That made no sense, either, and it got twisted in a number of ways; to "benjamin," for instance (because that, at least, was a familiar word, to "benjoin," and then finally, about 1650, to "benzoin." In English, the resin is now called "gum benzoin."

About 1608 an acid substance was isolated from the resin and that was eventually called "benzoic acid." Then, in 1834, a German chemist Eilhart Mitscherlich converted benzoic acid (which contains two oxygen atoms in its molecule) into a compound which contains no oxygen atoms at all, but only carbon and hydrogen atoms. He named the new compound "benzin," the first syllable signifying its ancestry.

Another German chemist Justus Liebig objected to the suffix -*in*, which, he said, was used only for compounds that contained nitrogen atoms, which Mitscherlich's "benzin" did not. In this, Liebig was correct. However, he suggested the suffix -*ol*, signifying the German word for "oil," because the compound mixed with oils rather than with water. This was as bad as -*in*, however, for, as I shall shortly explain, the suffix -*ol* is used for other purposes by chemists. However, the name caught on in Germany, where the compound is still referred to as "benzol."

In 1845 still another German chemist (I told you organic chemistry was a German monopoly in the nineteenth century) August W. von Hofmann suggested the name *benzene,* and this is the name properly used in most of the world, including the United States. I say properly, because the *-ene* ending is routinely used for many molecules containing hydrogen and carbon atoms only ("hydrocarbons") and therefore it is a good ending and a good name.

The molecule of benzene consists of six carbon atoms and six hydrogen atoms. The carbon atoms are arranged in a hexagon and to each of them is attached a single hydrogen atom. If we remember the actual structure we can content ourselves with stating that the formula of benzene is C_6H_6.

You will have noted, perhaps, that in the long and tortuous pathway from the island of Java to the molecule of benzene, the letters of the island have been completely lost. There is not a *j*, not an *a*, and not a *v*, in the word *benzene.*

Nevertheless, we've arrived somewhere. If you go back to the "Irish Washerwoman" compound, para-dimethyl-aminobenzaldehyde, you will not fail to note the syllable *benz.* Now you know where it comes from.

Having got this far, let's start on a different track altogether.

Women, being what they are (three cheers), have for many centuries been shading their eyelashes and upper eyelids and eye corners in order to make said eyes look large, dark, mysterious, and enticing. In ancient times they used for this purpose some dark pigment (an antimony compound, often) which was ground up into a fine powder. It had to be a *very* fine powder, of course, because lumpy shading would look awful.

The Arabs, with an admirable directness, referred to this cosmetic powder as "the finely divided powder." Only, once again, they used Arabic and it came out *"al-kuhl,"* where the *h* is pronounced in some sort of guttural way I can't imitate, and where *al* is the Arabic word for *the.*

The Arabs were the great alchemists of the early Middle Ages and when the Europeans took up alchemy in the late Middle Ages, they adopted many Arabic terms. The Arabs had begun to use *al-kuhl* as a name for any finely divided powder, without reference to cosmetic needs, and so did the Europeans. But they pronounced the word, and spelled it, in various ways that were climaxed with "alcohol."

As it happened, alchemists were never really at ease with gases or vapors. They didn't know what to make of them. They felt, somehow, that the vapors were not quite material in the same sense that liquids or solids were, and so they referred to the vapors as "spirits." They were particularly impressed with substances that gave off "spirits" even at ordinary temperatures (and not just when heated), and of these, the most important in medieval times was wine. So alchemists would speak of "spirits of wine" for the volatile component of wine (and we ourselves may speak of alcoholic beverages as "spirits," though we will also speak of "spirits of turpentine").

Then, too, when a liquid vaporizes it seems to powder away to nothing, so spirits also received the name *alcohol* and the alchemists would speak of "alcohol of wine." By the seventeenth century the word *alcohol* all by itself stood for the vapors given off by wine.

In the early nineteenth century the molecular structure of these vapors was determined. The molecule turned out to consist of two carbon atoms and an oxygen atom in a straight line. Three hydrogen atoms are attached to the first carbon, two hydrogen atoms to the second, and a single hydrogen atom to the oxygen. The formula can therefore be written as CH_3CH_2OH.

The hydrogen-oxygen group (-OH) is referred to in abbreviated form as a "hydroxyl group." Chemists began to discover numerous compounds in which a hydroxyl group is attached to a carbon atom, as it is in the alcohol of wine. All these compounds came to be referred to generally as alcohols, and each was given a special name of its own.

For instance, the alcohol of wine contains a group of two carbon atoms to which a total of five hydrogen atoms are attached. This same combination was discovered in a compound first isolated in 1540. This compound is even more easily vaporized than alcohol and the liquid disappears so quickly that it seems to be overwhelmingly eager to rise to its home in the high heavens. Aristotle had referred to the material making up the high heavens as "aether" (see Chapter 8), so in 1730 this easily vaporized material received the name *spiritus aethereus,* or, in

English, "ethereal spirits." This was eventually shortened to "ether."

The two-carbon-five-hydrogen group in ether (there were two of these in each ether molecule) was naturally called the "ethyl group," and since the alcohol of wine contained this group, it came to be called "ethyl alcohol" about 1850.

It came to pass, then, that chemists found it sufficient to give the name of a compound the suffix -ol to indicate that it was an alcohol, and possessed a hydroxyl group. That is the reason for the objection to *benzol* as a name for the compound C_6H_6. Benzene contains no hydroxyl group and is not an alcohol and should be called "benzene" and not "benzol." You hear?

It is possible to remove two hydrogen atoms from an alcohol, taking away the single hydrogen that is attached to the oxygen, and one of the hydrogens attached to the adjoining carbon. Instead of the molecule CH_3CH_2OH, you would have the molecule CH_3CHO.

Liebig (the man who had suggested the naughty word *benzol*) accomplished this task in 1835 and was the first actually to isolate CH_3CHO. Since the removal of hydrogen atoms is, naturally, a "dehydrogenation," what Liebig had was a dehydrogenated alcohol, and that's what he called it. Since he used Latin, however, the phrase was *alcohol dehydrogenatus*.

That is a rather long name for a simple compound, and chemists, being as human as the next fellow (honest!), have the tendency to shorten long names by leaving out syllables. Take the first syllable of *alcohol* and the first two syllables of *dehydrogenatus*, run the result together, and you have *aldehyde*.

Thus, the combination of a carbon, hydrogen, and oxygen atom (-CHO), which forms such a prominent portion of the molecule of dehydrogenated alcohol, came to be called the "aldehyde group," and any compound containing it came to be called an "aldehyde."

For instance, if we return to benzene, C_6H_6, and imagine one of its hydrogen atoms removed, and in its place a -CHO group inserted, we would have C_6H_5CHO and that compound would be "benzenealdehyde" or, to use the shortened form that is universally employed, "benzaldehyde."

Now let's move back in time again to the ancient

Egyptians. The patron god of the Egyptian city of Thebes on the Upper Nile was named Amen or Amun. When Thebes gained hegemony over Egypt, as it did during the eighteenth and nineteenth dynasties, the time of Egypt's greatest military power, Amen naturally gained hegemony over the Egyptian gods. He rated many temples, including one on an oasis in the North African desert, well to the west of the main center of Egyptian culture. This one was well known to the Greeks and, later, to the Romans, who spelled the name of the god "Ammon."

Any desert area has a problem when it comes to finding fuel. One available fuel in North Africa is camel dung. The soot of the burning camel dung, which settled out on the walls and ceiling of the temple, contained white, salt-like crystals, which the Romans then called "sal ammoniac," meaning "salt of Ammon." (The expression "sal ammoniac" is still good pharmacist's jargon, but chemists call the substance "ammonium chloride" now.)

In 1774 an English chemist Joseph Priestly discovered that heating sal ammoniac produced a vapor with a pungent odor, and in 1782 the Swedish chemist Torbern Olof Bergmann suggested the name *ammonia* for this vapor. Three years later a French chemist, Claude Louis Berthollet, worked out the structure of the ammonia molecule. It consisted of a nitrogen atom to which three hydrogen atoms were attached, so that we can write it NH_3.

As time went on, chemists who were studying organic compounds (that is, compounds that contained carbon atoms) found that it often happened that a combination made up of a nitrogen atom and two hydrogen atoms ($-NH_2$) was attached to one of the carbon atoms in the organic molecule the resemblance of this combination to the ammonia molecule was clear, and by 1860 the $-NH_2$ group was being called an "amine group" to emphasize the similarity.

Well, then, if we go back to our benzaldehyde, C_6H_5CHO, and imagine a second hydrogen atom removed from the original benzene and in its place an amine group inserted, we would have $C_6M_4(CHO)(NH_2)$ and that would be "aminobenzaldehyde."

Earlier I talked about the alcohol of wine, CH_3CH_2OH, and said it was "ethyl alcohol." It can also be called (and frequently is) "grain alcohol" because it is obtained from the fermentation of grain. But, as I hinted, it is not the only alcohol; far from it. As far back as 1661, the English chemist Robert Boyle found that if he heated wood in the absence of air, he obtained vapors, some of which condensed into clear liquid.

In this liquid he detected a substance rather similar to ordinary alcohol, but not quite the same. (It is more easily evaporated than ordinary alcohol, and it is considerably more poisonous, to mention two quick differences.) This new alcohol was called "wood alcohol."

However, for a name really to sound properly authoritative in science, what is really wanted is something in Greek or Latin. The Greek word for wine is *methy* and the Greek word for wood is *yli*. To get "wine from wood" (i.e., "wood alcohol"), stick the two Greek words together and you have *methyl*. The first to do this was the Swedish chemist Jöns Jakob Berzelius, about 1835, and ever since then wood alcohol has been "methyl alcohol" to chemists.

The formula for methyl alcohol was worked out in 1834 by a French chemist named Jean Baptiste André Dumas (no relation to the novelist, as far as I know). It turned out to be simpler than that of ethyl alcohol and to contain but one carbon atom. The formula is written CH_3OH. For this reason, a grouping of one carbon atom and three hydrogen atoms ($-CH_3$) came to be referred to as a "methyl group."

The French chemist Charles Adolphe Wurtz (he was born in Alsace, which accounts for his Germanic name) discovered in 1849 that one of the two hydrogen atoms of the amine group could be replaced by a methyl group, so that the end product looked like this: $-NHCH_3$. This would naturally be a "methylamine group." If both hydrogen atoms were replaced by methyl groups, the formula would be $-N(CH_3)_2$ and we would have a "dimethylamine group." (The prefix *di-* is from the Greek *dis*, meaning "twice." The methyl group is added to the amine group twice, in other words.)

Now we can go back to our aminobenzaldehyde, $C_6H_4(CHO)(NH_2)$. If, instead of an amine group, we had used a dimethylamine group, the formula would be C_6H_4 (CHO) $(N(CH_3)_2)$ and the name would be "dimethylaminobenzaldehyde."

Let's think about benzene once again. Its molecule is a hexagon made up of six carbon atoms, each with a hydrogen atom attached. We have substituted an aldehyde group for one of the hydrogen atoms and a dimethylamine group for another, to form dimethylaminobenzaldehyde, but *which* two hydrogen atoms have we substituted?

In a perfectly symmetrical hexagon, such as that which is the molecule of benzene, there are only three ways in which you can choose two hydrogen atoms. You can take the hydrogen atoms of two adjoining carbon atoms; or you can take the hydrogens of two carbon atoms so selected that one untouched carbon-hydrogen combination lies between; or you can take them so that two untouched carbon-hydrogen combinations lie between.

If you number the carbon atoms of the hexagon in order, one through six, then the three possible combinations involve carbons 1,2; 1,3; and 1,4 respectively. If you draw a diagram for yourself (it is simple enough), you will see that no other combinations are possible. All the different combinations of two carbon atoms in the hexagon boil down to one or another of these three cases.

Chemists have evolved a special name for each combination. The 1,2 combination is *ortho* from a Greek word meaning "straight" or "correct," perhaps because it is the simplest in appearance, and what seems simple, seems correct.

The prefix *meta-* comes from a Greek word meaning "in the midst of," but it also has a secondary meaning, "next after." That makes it suitable for the 1,3 combination. You substitute the first carbon, leave the next untouched, and substitute the one "next after."

The prefix *para-* is from a Greek word meaning "beside" or "side by side." If you mark the 1,4 angles on a hexagon and turn it so that the 1 is at the extreme left, then the 4 will be at the extreme right. The two are indeed "side by side" and so *para-* is used for the 1,4 combination.

Now we know where we are. When we say "para-dimethylaminobenzaldehyde," we mean that the dimethylamine group and the aldehyde group are in the 1,4 relationship to each other. They are at opposite ends of the benzene ring and we can write the formula $CHOC_6H_4N(CH_3)_2$.

See?

Now that you know Gaelic, what do you suppose the following are?

1) alpha-dee-glucosido-beta-dee-fructofuranoside

2) two,three-dihydro-three-oxobenzisosulfonazole

3) delta - four - pregnene - seventeen - alpha, twenty - one, diol-three, eleven, twenty-trione

4) three-(four-amino-two-methylpyrimidyl-five-methyl)-four-methyl-five-beta-hydroxyethylthiazolium chloride hydrochloride

Just in case your Gaelic is still a little rusty, I will give you the answers. They are:

1) table sugar
2) saccharin
3) cortisone
4) vitamin B_1

Isn't it simple?

Part Four

BIOLOGY

13. The Lost Generation

THERE ARE DISADVANTAGES to every situation, however ideal it may seem. For instance, by extremely clever maneuvering, I have created the image of one who possesses universal knowledge. This, plus the possession of a magnetic glance, enables me to browbeat editors (present editor always excepted).

Having brought myself to this ideal pass, however, I find myself occasionally asked to speak on some subject far outside my field of competence. When I then protest (very feebly) that I know nothing about it, there is a loud, jovial laugh in response and a hearty slap on the shoulders and someone says, "Good old Asimov! Always joking!"

Well, I can't allow the destruction of the image, or I might starve to death, so I do the next best thing to knowing my subject; I cheat. I start talking about whatever it is I am supposed to be talking about and then I sneakily change the subject to something I know.

For instance, one July I found myself staring at an audience of a hundred and fifty specialists in "information retrieval," having agreed to give the featured talk of the evening. By spelling the words *information retrieval* I have just given you all the knowledge I possess on the subject. The talk I proceeded to give was off the cuff (as all my talks are) and is lost forever. However, the following is an approximation of parts of it, anyway.

A magic phrase these days is "information retrieval," the study of devices whereby knowledge, once found, need never be lost again.

So many are the busy minds in scientific research who are hacking away at the jungle of ignorance, so numerous and miscellaneous are the fragmented bits of knowledge so obtained, that keeping all of it safely in hand is a problem indeed.

The information is published in a myriad journals, digested and spewed out again in a thousand reviews, pounded into pulp and summarized in a variety of abstracts, then compressed into invisibility and recorded on miles of microfilm. The net result is that any one needle of information, even a most important and crucial one, found for a moment of time, is in constant danger of being lost, lost, lost in the haystacks upon haystacks that fill the shelves of our technical libraries.

To rescue an important bit of knowledge, to snatch it out of its dusty surroundings, shake it free of obscurement, and hold it up, gleaming, in the light of day, is the purpose of information retrieval. Librarians, scientists, cyberneticists, combine to devise new methods of indexing and crossfiling and hope to transfer the organized information into the colossal and unfailing memory of a computer, in order that at the touch of a punch-code, the machine might bring forth anything that is known on any subject that is desired.

Thus the devices produced by the advance of modern science will, it is hoped, correct the incapacity of the human mind to keep up with the advance of modern science.

Yet there remains a flaw in this self-corrective process of science, a flaw for which no one has yet proposed a remedy, and one for which no remedy may be possible. It is not enough, after all, to supply a scientist with the information he needs. Once the information is supplied, the scientist must be capable of looking at it and seeing its importance.

This may sound an easy thing to do, this looking at importance and seeing it, but it is not. In fact, it may well be the hardest thing in the world. It may require all the intuition and creative talent of the world's best minds to see how a single bit of a jigsaw puzzle may just complete a structure and turn a meaningless jumble of facts into a fruitful and beautiful theory.

Far from being able to rely on a machine for this particular purpose, we cannot even rely on men—except, perhaps, for a very few.

Imagine, for example, a crucial scientific discovery, one that completely revolutionizes a major branch of science and supplies elegant answers to key questions that have agitated scientists and philosophers for thousands of years. And suppose, further, that there is but one major flaw in the discovery; one weakness that threatens to make this beautiful discovery worth nothing after all. A continent of scientists is searching desperately for a method of removing the flaw and, behold, the necessary piece of information is unexpectedly discovered by an amateur, and is worked out in full detail, so that the great central theory is complete at last.

Consider, next, that this piece of information, this key, this crux, is carefully placed in the hand of one of the most eminent scientists of the day, one who is bending his every effort to discovering just this piece of information. Now he has "retrieved" it; he has it.

What do you suppose the scientist would do with this information?

There is no need to guess. All that I have just described actually happened a hundred years ago. And in real life, the scientist who came upon the key contemptuously threw it away and kept on looking (in vain) for that which he had had and had not recognized. No one else found that thrown-away item for thirty-four years!

The great theory to which I have referred is that of "evolution by natural selection" as advanced by the English naturalist Charles Robert Darwin. This he did in 1859, in his book *The Origin of Species*, undoubtedly the most important single scientific work in all the history of the life sciences.

Ever since the time of the Greek philosophers, there have been scholars who studied the nature of the various species of plants and animals and who came to feel (sometimes very uneasily) that there was an ordered relationship among all those species; that one species might develop out of another; that several species might have a common ancestry.

The great difficulty at first was that no such evolutionary process was visible in all the history of man, so that if it occurred at all, it must take place with exceeding slowness. As long as mankind believed the earth to be no more than a few thousand years old, evolution was an impossible concept.

In the early nineteenth century, however, the conviction grew and strengthened that the earth was not a few thousand years old, but many millions of years old, and suddenly there was time for evolution to take place after all.

But now another problem arose. *Why* should evolution take place? What force drove some primitive antelope-like creature to lengthen its legs and neck and become a giraffe (which, despite its grotesque shape, still plainly revealed, in its anatomy and physiology, its relationship to the world of antelopes). Or why should a primitive four-hoofed creature, no larger than a dog, grow over the eons and lose one toe after another until it became the large one-hoofed creature we know today as the horse.

The first man to supply a reason was the French naturalist Jean Baptiste de Lamarck. In 1809 he suggested that animals changed because they voluntarily tried to change. Thus, a primitive antelope that dined on the leaves of trees was apt to find the leaves within easy reach already consumed by himself and his confreres. He would therefore stretch his neck and his legs, and even his tongue, to grasp leaves that were just higher than he could comfortably reach. A lifetime of such exercise would permanently (it seemed to Lamarck) extend the stretched portions just a trifle and the young of such a creature would inherit this slightly increased length of neck and limbs. (This is the doctrine of "inheritance of acquired characteristics.") The new generation would repeat the process and, very slowly, with the passage of time, the antelope would become a long-legged, long-necked, long-tongued giraffe.

This theory foundered on two points. In the first place, no evidence existed that showed that acquired characteristics could be inherited. In fact, all the evidence that biologists could locate proved just the reverse, the acquired characteristics were *not* inherited.

Secondly, Lamarckism might be conceivable for characteristics that could be altered by conscious effort, but what about other characteristics? The giraffe had also developed the novelty of a blotched coat that caused it to blend into the spattered background of light and shadow under the sunlit trees upon whose leaves it fed. This protective coloration makes it easier for the giraffe to avoid the predatory gaze of the large carnivores. But how did the giraffe develop this specialized and un-antelope-like coloration? Surely it could not have tried to become blotchy and therefore have succeeded in becoming just a trifle blotchier in the course of its life and then have passed on that additional bit of blotch to its young.

It fell to Darwin to supply a better answer. He spent years worrying about evolution until one day he happened upon a book called *An Essay on the Principle of Population* by an English economist named Thomas Robert Malthus. In this book Malthus pointed out that the human population increases more rapidly than the food supply and that the population must therefore inevitably be kept down by famine, by the disease that accompanies under-nutrition, or by the wars fought by competing groups of human beings, each intent on salvaging for itself a more than fair share of the earth's limited food supply.

And if this held true for mankind, thought Darwin, why not for all living creatures on earth? Each species would multiply until it outran its food supply, and each would be cut down by hunger, disease, and by the activity of those who preyed upon it, until there was balance between the numbers of the species and the quantity of its food.

But when the species was winnowed out, which individuals would be eliminated? Why, on the whole, reasoned Darwin, those who were less well adjusted to the life they led. A species that fed by running down its prey would find that the slowest runners would be the first to starve. If the species avoided danger by hiding, those less capable of efficient concealment would be the first to be eaten. If all were subject to a particular parasite, those that were least resistant would be the first to sicken and die.

In this way the blind forces of nature would continually, in each generation, weed out the less well adapted and preserve the better adapted.

The giraffe would not *try* to lengthen its legs and neck, but those that were born with slightly longer legs and neck in the first place would eat better and survive longer —and have more young to inherit their own particular characteristics. In each generation the longest legs and neck would survive by "natural selection" and the inborn length (*not* the acquired length) would be inherited.

Again, a giraffe that happened to be born with a blotchier pelt than average would more likely survive, so that the blotches would become more pronounced with the generations. It was not necessary for a giraffe to try for blotches. Natural selection would see to it.

And among the four-hoofed creatures ancestral to the modern horse, those which were largest (hence strongest and fleetest) would most easily survive in each generation. And those with the strongest central hoof (and hence with the leg best adapted, mechanically, for speed) would best survive. In the end, the large one-hoofed horse would be developed.

Darwin's theory created a terrific furor, but the loudest objections were the least crucial, scientifically. Of course, evolution by natural selection offended the religious sensibilities of many men, since it seemed to deny the version of the creation story found in the first chapter of the Book of Genesis. That form of opposition was by far the most dramatic, and culminated in the Scopes trial in Tennessee in 1924. However, this form of opposition played no great role within the realm of science itself.

Among scientists who were ready to accept the fact of evolution, there were still many who were not ready to accept the Darwinian mechanism. Natural selection was a blind, random force and to many the thought that the crowning creation of man should be brought about by the unseeing stagger of chance was intellectually repugnant.

Yet within the decade after the publication of Darwin's book, random forces were shown to account for some subtle facts in physics and chemistry. All the physical-chemical properties of gases, for instance, were found to result from the random movements of molecules. Randomness proved respectable and this form of opposition weakened.

However, there remained one objection so insuperable that if it were allowed to stand, it would be the ruin of the entire Darwinian theory. The theory's supporters could only suppose (and hope) that eventually some way out would be found. This objection involved the manner in which variations—the longer neck, the blotchier coat, the stronger hoof—were preserved across the generations.

Darwin pointed out that the variations arose, in the first place, through sheer random effects. In every group of young, in every litter, in every set of seedlings, there were trifling variations; differences in size, color, and every other conceivable characteristic. It was upon these random variations that natural selection would seize.

But, how could such variations be passed on from generation to generation so that they would remain in being for a long enough time to allow the very slow workings of natural selection to bring about the necessary result? One could not count on a male giraffe with a longer-than-usual neck mating with a female giraffe with a longer-than-usual neck. He would be quite likely to mate with a female with an ordinary neck, but one which happened to be ready and waiting when the male giraffe was ready and willing.

In the same way, a large stallion might very well mate with a small mare; a well-fanged lion with a small-toothed lioness; an intelligent ape with a stupid one.

And what would happen if these unlikes mated? Darwin was a pigeon-fancier and he knew what happened when varieties of pigeon were crossbred. For that matter, everyone knows what happens when purebred varieties of domestic animals are allowed to mate at random. The result is the mongrel; a creature in which the special characteristics of the ancestral varieties are blended into an undistinguished mixture. The sharp blacks and whites turn into a muddy gray.

Well, then, if chance forces produced the variations at birth, other chance forces would see to it that, through indiscriminate mating, those variations would blend and mix and cancel out before natural selection could get in its work.

So the theory of natural selection was simply unworkable in the light of the knowledge of Darwin's time. It seemed that despite all natural selection could do, species must remain middle-of-the-road and unchanging over countless eons. There could be no evolution, then, and yet there indubitably seemed to be evolution.

Some way had to be found out of this dilemma; some

way of destroying the paradox. The theory of evolution by natural selection had to be equipped with a driving mechanism that would push it onward.

One of those most intent on finding this driving mechanism was a Swiss botanist named Karl Wilhelm von Nägeli, a professor at the University of Munich. He was heir to a nineteenth-century school of German biologists who called themselves "nature philosophers."

The nature philosophers were a group who believed in the mystic importance of the individual and in the existence of misty and undefined forces particularly associated with life. The German language is particularly well adapted to a kind of learned professorial prose that resembles a cryptogram to which no key exists, and the nature philosophers could use this sort of language perfectly. If obscurity is mistaken for profundity, then they were profound indeed.

Von Nägeli was a perfectly competent botanist as long as he confined himself to making observations and reporting on them. When, however, he theorized and attempted to construct vast realms of thought, he produced nothing of value. His books were as thunderous as so many drums, and as empty.

To find a driving mechanism for the Darwinian theory, he went old Lamarck one better and postulated a mysterious inner drive that forced a species onward in change.

In this way, von Nägeli could forget the matter of random mating and the blending of characteristics that resulted. In fact, he could forget all concrete evidence and all reality, for he had solved the matter of a driving mechanism by simply assuming that one existed, without ever realizing that he was arguing in a circle.

He maintained that if the individuals of a species started to grow larger with the generations, the unconscious drive within that species would force the individuals to continue to grow larger. It did that because it did that because it did that because it did that. In fact, according to von Nägeli, this process (which he called "orthogenesis") would force the species to continue to grow larger, even past the point of diminishing returns, so that its oversize would eventually harm it and drive it to extinction.

(Needless to say, no biologist has taken orthogenesis seriously for a long time now.)

155

Meanwhile, in Brünn, Austria (now Brno, Czechoslovakia), there lived an Augustinian monk named Gregor Johann Mendel. He was far removed from the violent controversy that was then racking the world of biology. He had two interests outside the religious life and these were botany and statistics. With commendable economy he combined the two by growing pea plants in the monastery garden, and counting the different varieties he produced.

There are certain advantages to growing pea plants. First, they are docile creatures who do not resist the interference of man. Mendel could fertilize them in any combination he chose and so could control their pattern of mating with ease. Secondly, he could cause a pea plant to fertilize itself so that he could simplify matters by dealing with only one parent, rather than two. Finally, he could study individual characteristics that were much simpler than the more noticeable characteristics of, let us say, domestic animals such as dogs and cattle.

In crossing his pea plants during the 1860s, Mendel came across a number of fascinating effects that proved of prime importance. I'll mention two of them. First, characteristics did *not* blend and mix. Black and white did *not* produce gray.

When he crossed pea plants producing green peas with those producing yellow peas, he found that all the seedlings that resulted eventually produced yellow peas. The peas were not some-yellow-some-green; they were not all yellowish-green. They were all yellow, as yellow as though no green-pea parent had been involved.

Secondly, Mendel discovered that although the green peas had apparently disappeared in the second generation when all the pea plants produced yellow peas, they reappeared again in the third generation. In that generation, some of the yellow-pea plants produced some seedlings capable of producing green peas and others capable of producing yellow peas.

The deductions drawn from these facts, and from others which Mendel uncovered, are today called the "Mendelian laws of inheritance," and in the century that has passed there has been no reason to change the fundamentals. As Mendel discovered them, so they have remained.

Nor do the Mendelian laws apply only to pea plants,

or only to the plant world. If varieties of dogs seem to mongrelize when intermated, it is because so many different characteristics are involved. The crossbred young will inherit some characteristics from one parent, and some from another, so that *as a whole* it will seem intermediate. Each individual characteristic is inherited intact, in one fashion or another.

The Mendelian laws of inheritance provided just the driving mechanism that the Darwinian theory needed. If a desirable characteristic turned up, it would hang on for generation after generation, and remain in full, undiluted force, as the yellow peas did. Even if the characteristic seemed to disappear for a while, as the green peas did, it was not dead but was merely hiding and in the fullness of time it would appear again.

The reasons for all this were not worked out for decades to come, but the *facts* were incontrovertible. Characteristics did *not* blend together into an undistinguished middleground as a result of random mating. Instead, even the most random mating did not affect the emergence of different characteristics, and upon those characteristics natural selection seized and exerted its force.

But now that Mendel had made his crucial discovery (and he himself by no means recognized its crucial nature, for he was no evolutionist), what was he to do with it?

As he was only an unknown amateur, he felt that the best he could do would be to send his findings to some renowned and near-by botanist. If that botanist were pleased, he could then lend his name and prestige to the paper and bring it before the attention of the world. So Mendel sent it to von Nägeli.

Von Nägeli now had the key finding in his hands. He was the most fortunate (or, at least, he could have been) of all biologists of his generation. Darwin knew about evolution by natural selection but knew nothing about the Mendelian laws of inheritance. Mendel knew about the laws of inheritance but was not concerned with evolution by natural selection.

Only von Nägeli, in all the world, was now in a position to consider both, put them together, and find the first truly workable theory of evolution.

Looking back on it now, that would seem a simple

thing to do, but von Nägeli did not do it. Instead, he read Mendel's paper with the utmost disdain. It was not merely that Mendel was an unknown and an amateur, it was also true that the paper was full of numbers and ratios in an age when biologists never dealt with mathematics.

Moreover, to nature philosophers like von Nägeli, the important job of a biologist was the manufacture of windy and abstruse theories. To content one's self with counting pea plants seemed an idle amusement that could only be childish at best and idiotic at worst.

Von Nägeli returned the paper to Mendel with a curt, cold comment to the effect that the contents were not reasonable. Poor Mendel was crushed. He published the paper, in 1866, in the *Proceedings of the Natural History Society of Brünn* (a perfectly respectable periodical but rather obscure and out-of-the-way), and there it remained unsponsored and unnoticed.

Mendel never returned to his botanical work. In part, this was due to his increasing girth. He became stout enough to make it difficult to do the stooping that plant cultivation requires. He also became abbot of the monastery and found himself engaged in complicated controversy with the Austrian government over questions of taxation. However, the crushing rebuff from von Nägeli surely helped sour the whole subject of botanical research for him.

Mendel died in 1884 without having any notion that in the future there would be such a thing as "Mendelian" laws. Darwin died in 1882 without ever realizing that the major flaw in his theory had been corrected. And von Nägeli died in 1891 never for one moment suspecting that he had had the pearl of great price in his hand, and had thrown it away.

Even as von Nägeli lay dying, however, a Dutch botanist Hugo de Vries was working on the concept that evolution proceeded by jumps, by sudden changes called "mutations."

De Vries uncovered plants in which new varieties had sprung up, seemingly from nowhere, and observed that these new varieties maintained themselves over the generations and did *not* blend in with the other more normal varieties with which they might be crossed.

By 1900 he had worked out the same laws of inheritance

that Mendel had. Unknown to de Vries and to each other, two other botanists, a German, Karl Correns, and an Austrian, Erich Tschermak, had reached the same conclusions in that same year.

All three botanists, before publishing their papers, looked through previous work on the subject (they should have done that first) and all three found Mendel's paper in the obscure journal in which it was buried.

It is one of the glories of scientific history that these three men, each of whom had independently made one of the greatest and most important discoveries in biology, immediately relinquished any thought of retaining credit. Each published his work as nothing more than confirmation of a discovery made by an unknown monk a generation earlier.

This was no small sacrifice, as can be shown by the results. Mendel, through the triple relinquishment of credit, is now immortally famous in the history of science and lends his name to the laws of inheritance. De Vries on the other hand, is famous to a considerably lesser degree, for the development of the mutation theory; and as for Correns and Tschermak, they are virtually unknown even to specialists.

We live now at a time in which the livest and most exciting branch of biology is that of genetics, that branch of science that deals with the inheritance of characteristics. Based originally on Mendel's findings it has broadened into a domain that overlaps the fields of physics and chemistry and now fills the major portion of the new science of "molecular biology."

The science of molecular biology holds the promise of solving, at last, some of the most fascinating problems of life, and knowledge in this field is moving so quickly that no one dares predict where we will be ten years from now.

Where would we be now, then, if there had not been the sad loss of a complete generation of effort between Mendel and de Vries? What if, for those thirty-four years, men of science had been thinking of genetic problems and studying them instead of wasting their time on orthogenesis and such-like trash?

To be sure, nineteenth-century techniques would not have advanced them very far in that interval, as viewed by present standards, but there would have been some advance, certainly, which would be reflected in a better position for ourselves now.

But that generation, alas, is lost, and regrets are useless.

And yet we have no reason to suppose it can't happen again. What pair of eyes is gazing right now at a crucial finding, without seeing its significance? What hands are putting it to one side and what mind is closing against it?

We can't tell. We can never tell.

We can only hope that when the marvels of information retrieval put the right item before a man, it is put before the *right* man. And for human retrieval, no theory and no machinery exists. We can only hope.

14. He's Not My Type

I SEEM TO BE A NONCONFORMIST. This is not by any means because I have deliberately set out to be one. On the contrary, nothing would suit me better than to fade into the surroundings. Unfortunately, it turns out that at any gathering I attend I seem, for some mysterious reason, to attract attention.

Sooner or later, some curious stranger is bound to ask, "Who is the loud-mouth extrovert over there?"[1] And someone else is bound to say, "That's Asimov," and accompany the information with several taps on the forehead, a gesture of whose significance I am uncertain.

In response to this, I am forced back on the mumbled defense that everyone is different and has his own peculiarities, so there. (It's either that or stop being a loud-mouthed whatchamacallit.)

And I'm not wrong either. The fact that everyone is different is known perfectly well to all of us. An infant quickly learns to tell his mother from other women and

---

[1] Actually, he says "loud-mouthed nut," but I think the word *extrovert* is more accurate and has a more literary ring to it.

a young woman is very likely to be considered by her young man to be not only different from all others, but infinitely superior to all the others put together. I am told that young women (with less reason, no doubt) have similar feelings with regard to specific young men.

But placing these intuitively-felt individual differences on a hard, scientific foundation had to await the turn of the present century. Only then was it indubitably established that there was blood and blood.

Throughout history men have attributed differences to blood—but all the wrong differences. There was masculine red blood and aristocratic blue blood; and people talked of blood lines when they meant generations of a family. They spoke of good blood and of bad blood in the moral sense rather than the physical one, so that if you said of a person "He has bad blood," you didn't mean he had leukemia, but that his father had once forged a check. "It's in the blood," people would say meaningfully.

When the actual differences among blood were discovered, it turned out to be a very prosaic matter. It had nothing to do with morals or temperament or one's place in life. It was just that blood from one person didn't always mix well with blood from another.

The consequences of this fact had been apparent for centuries, actually. When someone was near death from loss of blood, it didn't take much imagination to decide that a little blood transferred into the patient's veins from another person in the full flush of health (and therefore able to spare a little blood) might stave off death. Occasionally doctors tried this, and occasionally the patient recovered. But occasionally the patient died almost at once.

The deaths were horrifying, of course, and doctors were forbidden by most enlightened governments to attempt transfusions.

In 1900, however, the matter was finally rationalized by an Austrian physician named Karl Landsteiner. He experimented by mixing red blood corpuscles from the blood of one individual with serum[2] from the blood of another.

---

[2] The liquid portion of blood is called "plasma." If a protein clotting factor, fibrinogen, is removed from plasma, what is left is serum. In practical matters, the two terms are virtually interchangeable.

In some cases, nothing happened. The corpuscles distributed themselves happily through the foreign serum and all was well. In other cases, however, the corpuscles, upon being added to the serum, adhered to each other in clumps. They had "agglutinated."

Clearly, then, there were at least two kinds of corpuscles and it seemed reasonable to suppose that the difference was chemical. One variety of corpuscle contained a chemical which, in the presence of the serum, reacted in such a way as to give rise to agglutination.

If we call this chemical "A" (you can't be simpler than that), then we can suppose there is a substance in the serum that reacts with it and we can call the serum-substance "anti-A."

Using this terminology we can say that if we have serum containing anti-A, we expect A corpuscles to agglutinate and other corpuscles not to agglutinate.

But this isn't all. It is also possible to obtain samples of serum from particular people that will *not* agglutinate A corpuscles but that *will* agglutinate corpuscles left untouched by anti-A. There must then be a second chemical present in some corpuscles, one which we can call (you guessed it) "B," and there must be varieties of serum that contain "anti-B."

We can now say that serum which contains anti-A will agglutinate A corpuscles but not B corpuscles, while serum which contains anti-B will agglutinate B corpuscles but not A corpuscles.

And still this isn't all. There are samples of red blood corpuscles which will agglutinate in both types of serum and which therefore contain *both* A and B. We can refer to these as AB corpuscles. Finally, there are red blood corpuscles which will agglutinate in neither type of serum and which therefore contain *neither* A nor B. These are O corpuscles ("oh," that is, and not "zero").

Every person, then, belongs to one of four "blood groups" or "blood types," depending on whether his red blood corpuscles contain A, B, both A and B, or neither A nor B. Furthermore, tests show that each person contained those antisubstances in his serum which would *not* react with his own corpuscles. (Obviously, or he would be dead to begin with.)

We can prepare a small table then:

Blood Type	Corpuscles	Serum
O	—	anti-A, anti-B
A	A	anti-B
B	B	anti-A
AB	A, B	—

By keeping a supply of serums containing known anti-A and anti-B, any sample of blood can be quickly typed, and transfusion can then be made safe. Transfusion is possible, without complications, when donor and patient are of the same blood type. No agglutination takes place and the donated blood flows freely through the patient's blood vessels.

Things are not necessarily ruinous even when donor and patient are of different types.

To explain that, let's begin by supposing that the blood of a B donor is given to an A patient. The donated blood is, roughly, half corpuscles and half serum and each half is a source of possible trouble.

The serum of the B donor contains anti-A which could bring about the agglutination of the patient's A corpuscles. This is not particularly serious. The half pint of serum donated by the B donor does not contain enough anti-A to do much damage, especially when it is quickly diluted by several quarts of the patient's own blood.

The second possibility is that the donor's B corpuscles may be agglutinated by the anti-B in the patient's serum. This is the real danger because it is the anti-B in an entire blood stream that must now be considered. If the corpuscles of the donated blood agglutinate, they are virtually useless for the performance of their chief function, that of transporting oxygen. Worse than that, the clumps of corpuscles will swirl through the blood stream, plugging tiny arteries in the kidney and elsewhere, and this is very likely to kill the patient.

In considering transfusion dangers, then, it is important to check the donor's red cells (not serum) and the patient's serum (not red cells).

Begin with an AB donor. His AB corpuscles cannot safely be given to any patient with either anti-A or anti-B in his serum. This means, if you look at the table above, that AB blood can be given *only* to an AB patient.

A sample of A blood can be donated only to patients without anti-A in the serum, which means that it can be given to either A or AB patients. Similarly a sample of B blood can be given to either B or AB patients. People with O blood have corpuscles that will not agglutinate in the presence of either anti-A or anti-B and such blood can be given to anyone. People of blood type O are therefore sometimes called "universal donors."[3]

This can be summarized in the following table:

Donor	Patient
AB	AB
A	AB, A
B	AB, B
O	AB, A, B, O

When much blood is needed for transfusions, as during wars, or even during lesser catastrophes, blood type O is particularly desirable.

This reminds me, always, of an occasion during World War II when I had given blood and was sitting at the Red Cross center with a glass of milk and a cookie, recovering from the ordeal. A loudmouthed extrovert sitting nearby was also recuperating and he announced himself to be of blood type O. I looked up and could see at once that he was not my type, for I am a B.

Someone asked the fellow why people of blood type O were so desired at the blood banks, and the fellow replied, with an insufferable smugness I found very difficult to take, "Well, O blood is particularly *rich,* you see."

Fortunately, I recover from these blows to my pride quite quickly. I've been brooding about this one for only sixteen years and expect to get over it fairly soon.

Anyway, Landsteiner's discovery made transfusion safe, snatched uncounted numbers of lives out of the jaws of

---

[3]This term is actually a slight exaggeration. Sometimes the anti-substance concentration in O blood is too high for comfort and wreaks a bit of havoc among the patient's corpuscles. Consequently, the practice of having donor and patient of the same blood type whenever possible is safest. It is also possible on occasion to do good by transfusing only plasma, eliminating the red blood corpuscles and with them virtually all the danger of transfusion.

death, and, as a result, it took only a full generation for the powers that be to decide he deserved a Nobel Prize in Medicine. He received it in 1930.

For purposes of transfusion there are four types of human blood, but the number is greater from the genetic viewpoint. Every person inherits two genes governing the particular blood groups I have been discussing, one from his mother and one from his father. Each gene can bring about the production of A, or B, or of neither, so that the genes are spoken of as belonging to the A, B, O group.

You can inherit any of six possible combinations then: OO, AO, AA, BO, BB, AB. When you possess the AO combination, the one A gene brings about the production of A corpuscles just as well as two A genes would. You are of blood type A, then, whether your combination is AA or AO. By similar reasoning, you are of blood type B, whether your combination is BB or BO. Your gene combination is your "genotype" and what you actually appear to be by test is your "phenotype." In other words, the six possible genotypes work out to four phenotypes.

But, you may ask, what does it matter whether you are AO or AA? Your blood reacts equally in either case, so why make a point of it? As far as transfusion goes, to be sure, the difference is negligible. But consider—

If two AA individuals marry, each can contribute only A genes to their offspring. All their offspring *must* be of blood type A. On the other hand, if two AO individuals marry, then it is possible that each will contribute an O gene to a particular offspring, which will then be OO and will test out as blood type O.

In other words, if two people, both of blood type A, marry, it is possible for an offspring to be of blood type O, without any hanky-panky having been involved. The existence of the AO genotype as opposed to the AA genotype is thus very important in paternity suits.

It was eventually found that there were two kinds of A corpuscles, one that reacted strongly with anti-A, and one that reacted weakly. The former was called A_1 and the latter A_2. This difference is of little importance in transfusion, but is, again, significant in paternity suits, since, for example, two A_1 parents cannot have an A_2 child, and vice versa.

Counting the two A varieties, we have ten genotypes, which I won't bother to list, giving rise to six phenotypes:

$$O, A_1, A_2, B, A_1B, \text{ and } A_2B$$

The reason why the A, B, O group of substances in the corpuscles was discovered as early as it was, rests with the fact that blood serum contains antisubstances that react with appropriate corpuscles and agglutinate them. But what if the corpuscles also contain other substances capable of bringing about agglutination which, however, do not make their presence felt, owing to the fact that the blood serum lacks the appropriate antisubstance?

If this were so, the only way of demonstrating the fact would be to produce the corresponding antisubstance artificially. This can be done by making use of the natural mechanisms of the animal body.

The body reacts to the injection of foreign proteins (and of certain other substances all lumped under the heading of "antigen") by producing an "antibody" which reacts with that antigen, removing it from circulation and rendering it harmless. Such a reaction is highly specific; that is, the antibody will react with the antigen and will react only weakly if at all with any other substance. Serum obtained from such a sensitized blood stream can then be used to detect the presence of this particular antigen through some sort of precipitating or clumping reaction.

In 1927 Landsteiner was able to show that rabbit blood could be sensitized in such a fashion that it would agglutinate some human corpuscles and not others, without reference to the A, B, O system. That is, some A corpuscles would be agglutinated but some not; some B corpuscles would be agglutinated and some not; and so on.

The obvious deduction was that there were additional corpuscle substances that were inherited independently of the A, B, O groups. These were labeled M and N, and any individual could be of blood type M, of blood type N, or of blood type MN. Serums containing anti-M and anti-N could be obtained from properly sensitized rabbits, and the human blood types could then be determined by noting whether corpuscles were agglutinated by anti-M, by anti-N, or by both.

This triples the number of phenotypes, for a person who is of blood group O can check out as blood group

OM, ON, or OMN. The analogous situation is true for the other blood groups. Out of the six genes, O, A_1, A_2, B, M, and N, then, eighteen phenotypes are possible.

In 1940 Alexander S. Wiener, an American physician, discovered that when rabbit blood was sensitized against red blood corpuscles obtained from a Rhesus monkey, the rabbit's serum could then be used to distinguish among blood from different human beings in still another fashion.

Apparently then, the blood corpuscles contain substances that belong neither to the A, B, O group nor to the M, N group. The new substances are referred to as the "Rh group," "Rh" standing for Rhesus monkey.

I hesitate to try to explain the ins and outs of the various Rh groups, because for twenty years now there has been a fairly violent running fight between various groups of immunologists as to just how to explain those same ins and outs—and I do not wish to get involved in it.

Apparently, though, there are at least twelve different Rh phenotypes that can be detected by using four different antisubstances. The three best known of the antisubstances are called anti-C, anti-D, and anti-E by some of the people in the field.

One of the phenotypes can be detected by the fact that the red blood corpuscles do not agglutinate in response to any of these three antisubstances and this phenotype is called "Rh-negative." All the other phenotypes agglutinate in response to one or another (in some cases, to more than one) of these antisubstances, and all eleven are lumped together under the general heading of "Rh-positive."

This turns out to be of importance not to transfusion, but in childbirth. When an Rh-negative mother is married to an Rh-positive father, the child may inherit, from the father, one of the Rh genes which will make it Rh-positive. This fact becomes true at the moment of conception and manifests itself during embryonic life. The situation then arises of an Rh-negative mother carrying an Rh-positive fetus.

The Rh-positive substances of the fetal corpuscles may make their way across the placental barrier into the maternal blood stream. The mother manufactures an antisubstance in response (since these Rh-positive substances do not naturally occur in her own blood). This antisubstance may then make its way back across the placental barrier into the fetal blood stream. The poor fetus now has both the substance and the antisubstance in the blood and is, so to speak, allergic to itself. If it is not stillborn altogether, it is born very sick with a condition called "erythroblastosis fetalis." It is usually fatal unless extensive transfusion is arranged for at once in order to remove the troublesome antisubstance.

The situation does not always arise, of course, and it almost never arises at the first pregnancy. It is estimated that about one birth out of four hundred in the United States involves erythroblastosis fetalis. Still, doctors like to be ready, just in case, which is why pregnant women are routinely typed for the Rh groups.

In any case, if we consider the twelve Rh phenotypes, we can see that each of the eighteen phenotypes involving the A, B, O and the M, N groups can be subdivided into twelve classes, one for each of the Rh phenotypes. The total number of blood types involving these three groups is therefore eighteen times twelve, or 216.

These various phenotypes are not, of course, evenly distributed. In the United States, for instance, 45 percent of the population is of blood type O, 42 percent of blood type A, 10 percent of blood type B, and 3 percent of blood type AB.

This distribution is American but not world-wide. There are American Indian tribes that are 98 percent O and 2 percent A, while other American Indian tribes are 80 percent A and 20 percent O. Practically no American Indians are B or AB.

The usual explanation for this is that the American Indians are descended from small groups of individuals who made their way across Siberia, over the Bering Strait, and down the American continent. The individuals who made it happened not to include any B types. (Since B is considerably less common than either A or O in the world as a whole, it is the more easily "lost" in small groups.) Alternatively, the comparatively few B individuals that reached America happened to die out without establishing a family line.

This loss of a particular gene among small groups is called "genetic drift."

On the other hand, blood type B, while always in a

minority, is most strongly represented (up to 30 percent) in Central Asia. Its frequency declines as one travels westward. It is down to 20 percent on the European border, 15 percent in western Russia, 10 percent in Germany, and 5 percent in France. Some people suggest that the B gene was brought into Europe by successive floods of Asian invaders, notably the Huns and Mongols.

In fact, there are attempts made to follow human migrations by tracing the variations in gene frequencies. These, however, are not always easy to work out and modern means of transportation are so churning up the human race that any remaining trace will, it seems to me, shortly be wiped out.

Anthropologists also try to work out a division of the human species into smaller groups on the basis of gene frequencies. For instance, the American Indians and the Australian aborigines are both marked by lack of the B gene. However, the American Indians are unusually high in M and low in N, whereas the Australian aborigines are unusually high in N and low in M.

Again, Asian individuals of blood type A are almost exclusively A_1, while in Europe and Africa both A_1 and A_2 are strongly represented among such individuals. As another example, there is one Rh gene that seems to occur almost exclusively in Africa.

The most interesting result obtained by such subdivisions-by-blood-group-frequencies involves the Rh series. People native to the Americas, to Asia, to Australia, and to Africa are virtually never Rh-negative. Where Rh-negative does occur, it almost always turns out that there are European natives among the ancestors of the individual.

It is Europe, then, which is the great reservoir of Rh-negativity. Among Europeans and their descendants on the other continents (including the Americans, of course), one out of seven individuals is Rh-negative.

How does this happen? Are there any areas in Europe which are focal points for Rh-negative genes, as the Mongols of Central Asia seem to have been the focal point for B. The answer is yes, for there is a small group of people in northern Spain called the Basques.[4] Among the Basques,

---

[4] At this point, I am tempted to maneuver the chapter in such a way that I can casually refer to "putting all my Basques in one exit," but I shall refrain.

one out of three is Rh-negative, and nowhere else in the world is there so high a concentration of this phenotype.

It would seem, then, that the Basques represent the remaining remnant of a group of Rh-negative "Early Europeans" who were flooded out by the invasions of the Rh-positive "Indo-European peoples" who now populate Europe. In the mountain fastnesses of Europe's far west they managed to retain a last grip.

This possibility is made the more attractive by the fact that the Basque language is not Indo-European in nature and, in fact, has no known relationship to any other language, living or dead. (So confusing is the Basque language to people speaking the common European tongues that there is a legend that the devil has no power over the Basques. He cannot tempt them because, try as he might, with all his diabolical power, he cannot learn their language.)

Nor have new blood groups come to an end with 1940 and the Rh series. Animals continue to be sensitized in various ways and to produce serums that can, in turn, be used to type individuals in new fashions. New blood types with names such as Duffy, Kell, Kidd, Lewis, and Lutheran (usually named after the patients in whose blood they were first located) are constantly being reported.

As of now, about sixty different blood-type series are known. Some of them are uncommon, of course, and no one serological laboratory is equipped to classify human blood in each of the sixty series. (The best laboratories can handle about twenty, I think.)

It has been calculated that the number of phenotypes that could actually be differentiated by the proper serums, if all were available, would come to (hold your breath now!) no less than 1,152,900,000,000,000,000, or a little over one quintillion.

This number is 400,000,000 times the population of the earth, so that it is highly unlikely that any two people (barring identical twins) are of absolutely identical blood type. In fact, it is easily conceivable that no two human beings who ever lived (barring identical twins) were of absolutely identical blood type. Not only is he not your type; no one is anybody's type, most likely.

And this just involves blood and blood corpuscles. Undoubtedly other tissues of the body differ from individual to individual just as complicatedly; so do the dietary requirements, protein structure, metabolic details, and so on. Not one of us conforms completely.

And that explains why it is perfectly all right for me to be a loud-mouthed extrovert.

—I think.

Part Five

ASTRONOMY

15. The Shape of Things

EVERY CHILD comes staggering out of grammar school with a load of misstatements of fact firmly planted in his head. He may forget, for instance, as the years drift by, that the Battle of Waterloo was fought in 1815 or that seven times six is forty-two; but he will never, never forget, while he draws breath, that Columbus proved the world was round.

And, of course, Columbus proved no such thing. What Columbus did prove was that it doesn't matter how wrong you are, as long as you're lucky.

The fact that the earth is spherical in shape was first suggested in the sixth century B.C. by various Greek philosophers. Some believed it out of sheer mysticism, the reasoning being that the sphere was the perfect solid and that therefore the earth was a sphere. To us, the premise is dubious and the conclusion a *non sequitur,* but to the Greeks it carried weight.

However, not all Greek philosophers were mystics and there were rational reasons for believing the earth to be spherical. These were capably summarized by Aristotle in the fourth century B.C. and turned out to be three in number:

1) If the earth were flat, then all the stars visible from one point on the earth's surface would be visible from all other points (barring minor distortions due to perspective

and, of course, the obscuring of parts of the horizon by mountains). However, as travelers went southward, some stars disappeared beyond the northern horizon, while new stars appeared above the southern horizon. This proved the earth was not flat but had some sort of curved shape. Once that was allowed, one could reason further that all things fell toward earth's center and got as close to it as they could. That solid shape in which the total distance of all parts from the center is a minimum is a sphere, Q.E.D.

2) Ships on leaving harbor and sailing off into the open sea seemed to sink lower and lower in the water, until at the horizon only the tops were visible. The most reasonable conclusion was that the water surface, though it seemed flat, was a gently curving hill behind which the ships disappeared. Furthermore, since this effect was equally intense whatever the direction in which the ship sailed, the gently curving hill of the ocean seemed to curve equally in all directions. The only solid shape that curves equally in all directions is a sphere, Q.E.D.

3) It was accepted by the Greek philosophers that the moon is eclipsed when it enters the earth's shadow. As darkness crossed over the face of the moon, the encroaching shadow marked off a projection of the shape of the earth, and that shadow was always the segment of a circle. It didn't matter whether the moon were high in the sky or at either horizon. The shadow was always circular. The only solid for which all projections are circular is a sphere, Q.E.D.

Now, Aristotle's reasoning carried conviction. All learned men throughout history who had access to Aristotle's books, accepted the sphericity of the earth. Even in the eighth century A.D., in the very depth of the Dark Ages, St. Bede (usually called "the Venerable Bede"), collecting what scraps of physical science were still remembered from Greek days, plainly stated the earth was a sphere. In the fourteenth century Dante's *Divine Comedy*, which advanced a detailed view of the orthodox astronomy of the day, presented the earth as spherical.

Consequently, there is no doubt that Columbus knew the earth was a sphere. But so did all other educated men in Europe.

In that case, what was Columbus's difficulty? He wanted to sail west from Europe and cross the Atlantic to Asia.

If the earth were spherical, this was theoretically possible, and if educated men all agreed with the premise and, therefore, with the conclusion, why the resistance to Columbus's scheme?

Well, to say the earth is a sphere is not enough. The question is—how large a sphere?

The first person to measure the circumference of the earth was a Greek astronomer, named Eratosthenes of Cyrene, and he did it without ever leaving home.

If the earth were a sphere, as Eratosthenes was certain it was, then the sun's rays should, at any one instant of time, strike different parts of the earth's surface at different angles. For instance, on June 21, the sun was just overhead at noon in the city of Syene, Egypt. In Alexandria, Egypt (where Eratosthenes lived), the sun was not quite overhead at that moment but made a small angle with the zenith.

Eratosthenes knew the distance between Alexandria and Syene, and it was simple geometry to calculate the curvature of the earth's surface that would account for the displacement of the sun. From that one could further calculate the radius and the circumference of the earth.

Eratosthenes worked out this circumference to be 25,000 miles in our modern units of length (or perhaps a little higher—the exact length in miles of the unit he used is uncertain) and this is just about right!

About 100 B.C., however, a Greek geographer named Posidonius of Apamea checked Eratosthenes' work and came out with a lower figure—a circumference of 18,000 miles.

This smaller figure may have seemed more comfortable to some Greeks, for it reduced the area of the unknown. If the larger figure were accepted, then the known world made up only about one sixth of the earth's surface area. If the smaller figure were accepted, the earth's surface area was reduced by half and the known world made up a third of the earth's surface area.

Now the Greek thinkers were much concerned with the unknown portions of the earth (which seemed as unattainable and mysterious to them as, until recently, the other side of the moon seemed to us) and they filled it with imaginary continents. To have less of it to worry about must have seemed a relief, and the Greek astronomer

Claudius Ptolemy, who lived about A.D. 150, was one of those who accepted Posidonius's figure.

It so happened that in the latter centuries of the Middle Ages, Ptolemy's books were as influential as Aristotle's, and if the fifteenth-century geographers accepted Aristotle's reasoning as to the sphericity of the earth, many of them also accepted Ptolemy's figure for its circumference.

An Italian geographer named Paolo Toscanelli was one of them. Since the extreme distance across Europe and Asia is some 13,000 miles (a piece of knowledge geographers had become acquainted with thanks to Marco Polo's voyages in the thirteenth century) and the total circumference was 18,000 miles or less, then one would have to travel westward from Spain no more than 5000 miles to reach "the Indies." In fact, since there were islands off the eastern coast of Asia, such as the Zipangu (Japan) spoken of by Marco Polo, the distance might be only 4000 miles or even less. Toscanelli drew a map in the 1470s showing this, picturing the Atlantic Ocean with Europe and Africa on one side and Asia, with its offshore islands, on the other.

Columbus obtained a copy of the map and some personal encouragement from Toscanelli and was an enthusiastic convert to the notion of reaching Asia by the westward route. All he needed now was government financing.

The most logical place to go for such financing was Portugal. In the fifteenth century many of Europe's luxuries (including spices, sugar, and silk) were available only by overland routes from the Far East, and the Turks who straddled the route charged all the traffic could bear in the way of middleman fees. Some alternate route was most desirable, and the Portuguese, who were at the extreme southeastern edge of Europe, conceived the notion of sailing around Africa and reaching the Far East by sea, outflanking the Turks altogether. Throughout the fourteenth century, then, the Portuguese had been sending out expedition after expedition, farther and farther down the African coast. (The Portuguese "African effort" was as difficult for those days as our "space effort" is for ours.)

In 1484, when Columbus appealed to John II of Portugal for financing, Portuguese expeditions had all but reached the southern tip of Africa (and in 1487 they were to do so).

The Portuguese, at the time, were the most experienced

navigators in Europe, and King John's geographers viewed with distrust the low figure for the circumference of the earth. If it turned out that the high figure, 25,000 miles, were correct, and if the total east-west stretch of Europe and Asia were 13,000 miles—then it followed, as the night the day, that a ship would have to sail 12,000 miles west from Portugal to reach Asia. No ship of that day could possibly make such an uninterrupted ocean voyage.

The Portuguese decision, therefore, was that the westward voyage was theoretically possible but, given the technology of the day, completely impractical. The geographers advised King John to continue work on Project Africa and to turn down the Italian dreamer. This was done.

Now, mind you, the Portuguese geographers were perfectly right. It *is* 12,000 miles from Portugal west to Asia, and no ship of the day could possibly have made such a voyage. The fact is that Columbus never did reach Asia by the western route, whereas the Portuguese voyagers succeeded, within thirteen years, in reaching Asia by the African route. As a result, tiny Portugal built a rich and far-flung empire, becoming the first of the great European colonialists. Enough of that empire has survived into the 1960s to permit them to be the last as well.

And what is the reward of the Portuguese geographers for proving to be right in every last particular? Why, schoolchildren are taught to sneer at them.

Columbus obtained the necessary financing from Spain in 1492. Spain had just taken the last Moslem strongholds on the Iberian Peninsula and, in the flush of victory, was reaching for some daring feat of navigation that would match the deeds of the Portuguese. (In the language of today, they needed an "ocean spectacular" to improve their "world image.") So they gave Columbus three foundering hulks and let him have his pick of the prison population for crewmen and sent him off.

It would have meant absolutely certain death for Columbus and his men, thanks to his wrongness, were it not for his incredible luck. The Greek dreamers had been right. The unoccupied wastes of the earth did indeed possess other continents and Columbus ran aground on them after only 3000 miles. (As it was, he barely made it; another thousand miles and he would have been gone.)

The Portuguese geographers had not counted on what

are now known as the American continents (they would have been fools to do so), but neither had Columbus. In fact, Columbus never admitted he had reached anything but Asia. He died in 1506 still convinced the earth was 18,000 miles in circumference—stubbornly wrong to the end.

So Columbus had not proved the earth was round; that was already known. In fact, since he had expected to reach Asia and had failed, his voyage was an argument *against* the sphericity of the earth.

In 1519, however, five ships set sail from Spain under Ferdinand Magellan (a Portuguese navigator in the pay of Spain), with the intention of completing Columbus's job and reaching Asia, and then continuing on back to Spain. Such an expedition was as difficult for its day as orbiting a man is for ours. The expedition took three years and made it by an inch. An uninterrupted 10,000-mile trip across the Pacific all but finished them (and they were far better prepared than Columbus had been). Magellan himself died en route. However, the one ship that returned brought back a large enough cargo of spices to pay for the entire expedition with plenty left over.

This first circumnavigation of the earth was experimental confirmation, in a way, of the sphericity of the planet, but that was scarcely needed. More important, it proved two other things. It proved the ocean was continuous; that there was one great sea in which the continents were set as large islands. This meant that any seacoast could be reached from any other seacoast, which was vital knowledge (and good news) for merchantmen. Secondly, it proved once and for all that Eratosthenes was right and that the circumference of the earth was 25,000 miles.

And yet, after all, though the earth is round, it turned out, despite all Aristotle's arguments, that it wasn't a sphere after all.

Again we go back to the Greeks. The stars wheel about the earth in a stately and smooth twenty-four-hour cycle. The Greek philosophers realized that this could be explained in either of two ways. It was possible that the earth stood still and the heavens rotated about it in a twenty-four-hour period. Or the heavens might stand still while the earth rotated about itself in twenty-four hours.

A few Greeks (notably Aristarchus of Samos) maintained, in the third century B.C., that it was the earth that rotated. The majority, however, held for a stationary earth, and it was the latter who won out. After all, the earth is large and massive, while the heavens are light and airy; surely it is more logical to suppose the latter turned.

The notion of the stationary earth was accepted by Ptolemy and therefore by the medieval scholars and by the Church. It was not until 1543, a generation after Magellan's voyage, that a major onslaught was made against the view.

In that year Nicolaus Copernicus, a Polish astronomer, published his views of the universe and died at once, ducking all controversy. According to his views (which were like those of Aristarchus) the sun was the center of the universe, and the earth revolved about it as one planet among many. If the earth were only a minor body circling the sun, it seemed completely illogical to suppose that the stars revolved about our planet. Copernicus therefore maintained that the earth rotated on its axis.

The Copernican view was not, of course, accepted at once, and the world of scholarship argued the matter for a century. As late as 1633, Galileo was forced by the Inquisition to abjure his belief that the earth moved and to affirm that it was motionless. However, that was the dying gasp of the motionless-earth view, and there has been no scientific opposition to earth's rotation since. (Nevertheless, it was not until 1851 that the earth's rotation was actually confirmed by experiment, but that is another story.)

Now if the earth rotated, the theory that it was spherical in shape suddenly became untenable. The man who first pointed this out was Isaac Newton, in the 1680s.

If the earth were stationary, gravitational forces would force it into spherical shape (minimum total distance from the center) even if it were not spherical to begin with. If the earth rotated, however, the possession of inertia by every particle on the planet would produce a centrifugal effect, which would act as though to counter gravity and move particles *away* from the center of the earth.

But the surface of a rotating sphere moves at varying velocities depending upon its distance from the axis of rotation. At the point where the axis of rotation intersects the surface (as at the North and South Poles) the surface is motionless. As distance from the Poles increases, the surface velocity increases; it is at its maximum at the Equator, which is equidistant from the Poles.

Whereas the gravitational force is constant (just about) at all points on earth's surface, the centrifugal effect increases rapidly with surface velocity. As a result the surface of the earth lifts up slightly away from the center and the lifting is at its maximum at the Equator where the surface velocity is highest. In other words, said Newton, the earth should have an equatorial bulge. (Or, to put it another way, it should be flattened at the Poles.)

This means that if an east-west cross-section of the earth were taken at the Equator, that cross-section would have a circular boundary. If, however, a cross-section were taken north-south through the Poles, that cross-section would have an elliptical boundary and the shortest diameter of the ellipse would run from Pole to Pole. Such a solid body is not a sphere but an "oblate spheroid."

To be sure, the ellipticity of the north-south cross-section is so small that it is invisible to the naked eye and, viewed from space, the earth would seem a sphere. Nevertheless, the deviation from perfect sphericity is important, as I shall explain shortly.

Newton was arguing entirely from theory, of course, but it seemed to him he had experimental evidence as well. In 1673 a French scientific expedition in French Guiana found that the pendulum of their clock, which beat out perfect seconds in Paris, was moving slightly slower in their tropical headquarters—as compared with the steady motion of the stars. This could only mean that the force of gravity (which was what powered the swinging pendulum) was slightly weaker in French Guiana than in Paris.

This would be understandable if the scientific expedition were on a high mountain where the distance from the center of the earth were greater than at sea level and the gravitational force consequently weakened—but the expedition *was* at sea level. Newton, however, maintained that, in a manner of speaking, the expedition was not truly at sea level, but was high up on the equatorial bulge and that that accounted for the slowing of the pendulum.

In this, Newton found himself in conflict with an Italian-born French astronomer named Jean Dominique Cassini.

The latter tackled the problem from another direction. If the earth were not a true sphere, then the curvature of its surface ought to vary from point to point. (A sphere is the only solid that has equal curvature everywhere on its surface.) By triangulation methods, measuring the lengths of the sides and the size of the angles of triangles drawn over large areas of earth's surface, one could determine the gentle curvature of that surface. If the earth were truly an oblate spheroid, then this curvature ought to decrease as one approached either Pole.

Cassini had conducted triangulation measurements in the north and south of France and decided that the surface curvature was less, not in the north, but in the south. Therefore, he maintained, the earth bulged at the Poles and was flattened at the Equator. If one took a cross-section of the earth through the Poles, it would have an elliptical boundary indeed, but the longest (and not the shortest) diameter would be through the Poles. Such a solid is a "prolate spheroid."

For a generation, the argument raged. It was not just a matter of pure science either. I said the deviation of the earth's shape from the spherical was important, despite the smallness of the deviation, and that was because ocean voyages had become commonplace in the eighteenth century. European nations were squabbling over vast chunks of overseas real estate, and victory could go to the nation whose ships got less badly lost en route. To avoid getting lost one had to have accurate charts and such charts could not be drawn unless the exact deviation of the earth's shape from the spherical were known.

It was decided that the difference in curvature between northern and southern France was too small to decide the matter safely either way. Something more extreme was needed. In 1735, therefore, two French expeditions set out. One went to Peru, near the Equator. The other went to Lapland, near the North Pole. Both expeditions took years to make their measurements (and out of their difficulties arose a strong demand for a reform in standards of measurement that led, eventually, to the establishment of the metric system a half century later). When the expeditions returned, the matter was settled. Cassini was wrong, and Newton was right. The equatorial bulge is thirteen miles high, which means that a point at sea level on the Equator is thirteen miles farther from the center of the earth than is sea level at either Pole.

The existence of this equatorial bulge neatly explained one particular astronomic mystery. The heavens seem to rotate about an axis of which one end (the North Celestial Pole) is near the North Star. An ancient Greek astronomer, Hipparchus of Nicaea, was able to show about 150 B.C. that this celestial axis is not fixed. It marks out a circle in the heavens and takes some 25,800 years to complete one turn of the circle. This is called "the precession of the equinoxes."

To Hipparchus, it seemed that the heavenly sphere simply rotated slowly in that fashion. He didn't know why. When Copernicus advanced his theory, he had to say that the earth's axis wobbled in that fashion. He didn't know why, either.

Newton, however, pointed out that the moon traveled in an orbit that was not in the plane of the earth's Equator. During half of its revolution about the earth, it was well to the north of the Equator and during the other half it was well to the south. If the earth were perfectly spherical, the moon would attract it in an all-one-piece fashion from any point. As it was, the moon gave a special unsymmetrical yank at the equatorial bulge. Newton showed that this pull at the bulge produced the precession of the equinoxes. This could be shown experimentally by hanging a weight on the rim of a spinning gyroscope. The axis of the gyroscope then precesses.

And thus the moon itself came to the aid of scientists interested in the shape of things.

An artificial moon was to do the same, two and a half centuries after Newton's time.

The hero of the latest chapter in the drama of earth's shape is Vanguard I, which was launched by the United States on March 17, 1958. It was the fourth satellite placed in orbit and is currently the oldest satellite still orbiting and emitting signals. Its path carried it so high above earth's surface that in the absence of atmospheric interference it will stay in orbit for a couple of centuries. Furthermore, it has a solar battery which will keep it delivering signals for years.

The orbit of Vanguard I, like that of the moon itself, is not in the plane of the earth's Equator, so Vanguard I pulls on the equatorial bulge and is pulled by it, just as the moon does. Vanguard I isn't large enough to affect the earth's motion, of course, but it is itself affected by the pull of the bulge, much more than the moon is.

For one thing, Vanguard I is nearer to the bulge and is therefore affected more strongly. For another, what counts in some ways is the total number of revolutions made by a satellite. Vanguard I revolves about the earth in two and a quarter hours, which means that in a period of fourteen months, it has completed about 4500 revolutions. This is equal to the total number of revolutions that the moon has completed since the invention of the telescope. It follows that the motions of Vanguard I better reveal the fine structure of the bulge than the motions of the moon do.

Sure enough, John A. O'Keefe, by studying the orbital irregularities of Vanguard I, was able to show that the earth's equatorial bulge is not symmetrical. The satellite is yanked just a little harder when it is south of the Equator, so that the bulge must be a little bulgier there. It has been calculated that the southern half of the equatorial bulge is up to fifty feet (not miles but *feet!*) farther from the earth's center than the northern half is. To balance this, the South Pole (calculating from sea level) is one hundred feet closer to the center of the earth than the North Pole is.

So the earth is not an exact oblate spheroid, either. It is very, very, very slightly egg-shaped, with a bulging southern half and a narrow northern half; with a flattened southern tip and a pointy northern tip.

Nevertheless, to the naked eye, the earth is still a sphere, and don't you forget it.

This final tiny correction is important in a grisly way. Nowadays the national insanity of war requires that missiles not get lost en route, and missiles must be aimed far more accurately than ever a sailing vessel had to be. The exact shape of the earth is more than ever important.

Moreover, this final correction even has theoretical implications. To allow such an asymmetry in the bulge against the symmetrical pull of gravity and the push of centrifugal force, O'Keefe maintains, the interior of the earth must be considerably more rigid than geophysicists had thought.

One final word: O'Keefe's descriptive adjective for the shape of the earth, as revealed by Vanguard I, is "pear-shaped," and the newspapers took that up at once. The result is that readers of headlines must have the notion that the earth is shaped like a Bartlett pear, or a Bosc pear, which is ridiculous. There are some varieties of pears that are closer to the egg-shaped, but the best-known varieties are far off. However, "pear-shaped" will last, I am sure, and will do untold damage to the popular conception of the shape of the earth. Undoubtedly the next generation of kids will gain the firm conviction that Columbus proved the earth is shaped like a Bartlett pear.

But it is an ill wind that blows no good, and I am breathlessly awaiting a certain opportunity. You see, in 1960 a book of mine entitled *The Double Planet* was published. It is about the earth and moon, which are more nearly alike in size than any other planet-satellite combination in the solar system, so that the two may rightly be referred to as a "double planet."

Now someday, someone is going to pick up a copy of the book in my presence (I have my books strategically scattered about my house), and leaf through it and say, "Is this about the earth?"

With frantically beating heart, I will say, "Yes."

And he will say (I hope, I hope), "Why do you call the earth a double planet?"

And then I will say (get this now), *"Because it is pair-shaped!!!"*

—Why am I the only one laughing?

16. Twinkle, Twinkle, Little Star

IT CAME as a great shock to me, in childhood days, to learn that our sun was something called a "yellow dwarf" and that sophisticated people scorned it as a rather insignificant member of the Milky Way.

I had made the very natural assumption. prior to that, that stars were little things, and everything I had read confirmed the notion. There were innumerable fairy tales about the tiny stars, which (I gathered) must be the little children of the sun and moon, the brightly shining sun being the father and the dim, retiring moon the mother.[1]

When I found that all those minute points of light were huge, glaring suns greater than our own, it not only upset the sanctity of the heavenly family for me, but it also offended me as a patriotic inhabitant of the solar system. Consequently, it was with grim relief that I eventually learned that not all stars were greater than the sun after all; that, in fact, a great many were smaller than the sun.

What's more, I found some of those small stars to be intensely fascinating; and in order to talk about them, I will begin Asimov-fashion at the other end of the stick, and consider the earth and the sun.

The earth does not really revolve about the sun. Both earth and sun (taken by themselves) revolve about a common center of gravity. Naturally, the center of gravity is closer to the more massive body and the degree of closeness is proportional to the ratio of masses of the two bodies.

Thus, the sun is 333,400 times as massive as the earth, and the center of gravity should therefore be 333,400 times as close to the sun's center as to the earth's center. The distance between earth and sun, center to center, is about 92,870,000 miles; and dividing that by 333,400 gives us the figure 280. Therefore, the center of gravity of the earth-sun system is 280 miles from the center of the sun.

This means that as the earth moves around this center of gravity in its annual revolution, the sun makes a small circle 280 miles in radius about the same center, leaning always away from the earth. Of course, this trifling wobble is quite imperceptible from an observation point outside the solar system; say, from Alpha Centauri.

But what about the other planets? Each one of them revolves with the sun about a common center of gravity.

[1] I had curiously naïve ideas about the comparative importance of the sexes in those days.

Some of the planets are both more massive than the earth and more distant from the sun, each of these factors working to move the center of gravity farther from the sun's center. To show you the result, I have worked out the following table (which, by the way, I have never seen in any astronomy text).

Planet	Distance (miles) of Center of Gravity of Sun-Planet System from Center of Sun
Mercury	6
Venus	80
Earth	280
Mars	45
Jupiter	460,000
Saturn	250,000
Uranus	80,000
Neptune	140,000
Pluto	1,500 (?)

The radius of the sun is 432,200 miles, so the center of gravity in every case but one lies below the sun's surface. The exception is Jupiter. The center of gravity of the Jupiter-sun system is about 30,000 miles *above* the sun's surface (always in the direction of Jupiter, of course).

If the sun and Jupiter were all that existed in the solar system, an observer from Alpha Centauri, say, though not able to see Jupiter, might (in principle) be able to observe that the sun was making a tiny circle about something or other every twelve years. This "something or other" could only be the center of gravity of a system consisting of the sun and another body. If our observer had a rough idea of the mass of the sun, he could tell how distant the other body must be to impose a twelve-year revolution. From that distance, as compared with the radius of the circle the sun was making, he could deduce the mass of the other body. In this way, the observer on Alpha Centauri could discover the presence of Jupiter and work out its mass and its distance from the sun without ever actually seeing it.

Actually, though, the wobble on the sun imposed by Jupiter is still too small to detect from Alpha Centauri (assuming their instruments to be no better than ours). What makes it worse is that Saturn, Uranus, and Neptune (the other planets can be ignored) impose wobbles on the sun, too, which complicate its motion.

But suppose that circling the sun were a body considerably more massive than Jupiter. The sun would then make a much larger circle and a much simpler one, for the effect of other revolving bodies would be swamped by this super-Jupiter. To be sure, this is not the case with the sun, but is it possible that it might be so for other stars?

Yes, indeed, it *is* possible.

In 1834 the German astronomer Friedrich Wilhelm Bessel concluded, from a long series of careful observations, that Sirius was moving across the sky in a wavy line. This could best be explained by supposing that the center of gravity of Sirius and another body was moving in a straight line and that it was Sirius's revolution about the center of gravity (in a period of some fifty years) that produced the waviness.

Sirius, however, is about two and a half times as massive as the sun, and for it to be pulled as far out of line as observation showed it to be, the companion body had to be much more massive than Jupiter. In fact, it turned out to be about one thousand times as massive as Jupiter, or just about as massive as our sun. If we call Sirius itself "Sirius A," then this thousand-fold-Jupiter companion would be "Sirius B." (This use of letters has become a standard device for naming componets of a multiple star system.)

Anything as massive as the sun ought to be a star rather than a planet and yet, try as he might, Bessel could see nothing in the neighborhood of Sirius A where Sirius B ought to be. The seemingly natural conclusion was that Sirius B was a burned-out star, a blackened cinder that had used up its fuel. For a generation, astronomers spoke of Sirius's "dark companion."

In 1862, however, an American telescope-maker, Alvan Graham Clark, was testing a new eighteen-inch lens he had made. He turned it on Sirius to test the sharpness of the image it would produce, and, to his chagrin, found there was a flaw in his lens, for near Sirius was a sparkle of light that shouldn't be there. Fortunately, before going back to his grinding, he tried the lens on other stars, and the defect disappeared! Back to Sirius—and there was that sparkle of light again.

It couldn't be a defect; Clark had to be seeing a star. In fact, he was seeing Sirius's "dark companion," which wasn't quite dark after all, for it was of the eighth magnitude. Allowing for its distance, however, it was at least dim, if not dark, for it was only $\frac{1}{120}$ as luminous as our sun—there was still that much of a dim glow amid its supposed ashes.

In the latter half of the nineteenth century, spectroscopy came into its own. Particular spectral lines could be produced only at certain temperatures, so that from the spectrum of a star its surface temperature could be deduced. In 1915 the American astronomer Walter Sydney Adams managed to get the spectrum of Sirius B and was amazed to discover that it was not a dimly glowing cinder at all, but had a surface rather hotter than that of the sun!

But if Sirius B was hotter than the sun, why was it only $\frac{1}{120}$ as bright as the sun? The only way out seemed to be to assume that it was much smaller than the sun and had, therefore, a smaller radiating surface. In fact, to account for both its temperature and its low total luminosity, it had to have a diameter of about 30,000 miles. Sirius B, although a star, was just about the size of the planet Uranus.

It was more dwarfish than any astronomer had conceived a star might be and it was white-hot, too. Consequently, Sirius B and all other stars of that type came to be called "white dwarfs."

But Bessel's observation of the mass of Sirius B was still valid. It was still just about as massive as the sun. To squeeze all that mass into the volume of Uranus meant that the average density of Sirius B had to be 38,000 kilograms per cubic centimeter, or about 580 tons per cubic inch.

Twenty years earlier, this consequence of Adams' discovery would have seemed so ridiculous that the entire chain of reasoning would have been thrown out of court and the very concept of judging stellar temperatures by spectral lines would have come under serious doubt. By Adams' time, however, the internal structure of the atom had been worked out and it was known that virtually all the mass of the atoms was concentrated in a tiny nucleus at the very center of the atom. If the atom could be broken down and the central nuclei allowed to approach, the density of Sirius B—and, in fact, densities millions of times greater still—became conceivable.

Sirius B by no means represents a record either for the smallness of a star or for its density. Van Maanen's Star (named for its discoverer) has a diameter of only 6048 miles, so that it is smaller than the earth and not very much larger than Mars. It is one seventh as massive as our sun (about 140 times as massive as Jupiter), and that is enough to make it fifteen times as dense as Sirius B. A cubic inch of average material from Van Maanen's Star would weigh 8700 tons.

And even Van Maanen's Star isn't the smallest. In the course of this last year William J. Luyten of the University of Minnesota has discovered a white dwarf star with a diameter of about 1000 miles—only half that of the moon.

Of course, the white dwarfs can't really give us much satisfaction as "little stars." They may be dwarfs in volume but they are sun-size in mass, and giants in density and in intensity of gravitational fields. What about really little stars, in mass and temperature as well as in volume?

These are hard to find. When we look at the sky, we are automatically making a selection. We see all the large, bright stars for hundreds of light-years in all directions, but the dim stars we can barely see at all, even when they are fairly close.

Judging by the stars we see, our sun, sure enough, is a rather insignificant dwarf, but we can get a truer picture by confining ourselves to our own immediate neighborhood. That is the only portion of space through which we can make a reasonably full census of stars, dim ones and all.

Thus, within five parsecs (16½ light-years) of ourselves, according to a compilation prepared by Peter Van de Kamp of Swarthmore College, there are thirty-nine stellar systems, including our own sun. Of these, eight include two visible components and two include three visible components, so that there are fifty-one individual stars altogether.

Of these, exactly three stars are considerably brighter than our sun and these we can call "white giants."

Star	Distance (light-years)	Luminosity (sun = 1)
Sirius A	8.6	23
Altair	15.7	8.3
Procyon A	11.0	6.4

There are then a dozen stars that are as bright or nearly as bright as the sun. We can call these "yellow stars" without making any invidious judgments as to whether they are dwarfs or not.

Star	Distance (light-years)	Luminosity (sun = 1)
Alpha Centauri A	4.3	1.01
Sun	—	1.00
70 Ophiuchi A	16.4	0.40
Tau Ceti	11.2	0.33
Alpha Centauri B	4.3	0.30
Omicron$_2$ Eridani A	15.9	0.30
Epsilon Eridani	10.7	0.28
Epsilon Indi	11.2	0.13
70 Ophiuchi B	16.4	0.08
61 Cygni A	11.1	0.07
61 Cygni B	11.1	0.04
Groombridge 1618	14.1	0.04

Of the remaining stars, all of which are less than one twenty-fifth as luminous as the sun, four are white dwarfs:

Star	Distance (light-years)	Luminosity (sun = 1)
Sirius B	8.6	0.008
Omicron$_2$ Eridani B	15.9	0.004
Procyon B	11.0	0.0004
Van Maanen's Star	13.2	0.00016

This leaves thirty-two stars that are not only considerably dimmer than the sun, but considerably cooler, too, and therefore distinctly red in appearance. To be sure, there are cool red stars that are nevertheless much brighter in total luminosity than our sun because they are so gigantically voluminous. (This is the reverse of the white-dwarf situation.) These tremendous cool stars are "red giants," and there are none of these in the sun's vicinity—distant Betelgeuse and Antares are the best-known examples.

The cool, red, small stars are "red dwarfs." An example of this is the very nearest star to ourselves, the third and dimmest member of the Alpha Centauri system. It should be called Alpha Centauri C, but because of its nearness, it is more frequently called Proxima Centauri. It is only $\frac{1}{23,000}$ as bright as our sun and, despite its nearness, can be seen only with a good telescope.

To summarize, then, there are, in our vicinity: no red giants, three white giants, twelve yellow stars, four white dwarfs, and thirty-two red dwarfs. If we consider the immediate neighborhood of the sun to be a typical one (and we have no reason to think otherwise), then well over half the stars in the heavens are red dwarfs and considerably dimmer than the sun. Indeed, our sun is among the top 10 percent of the stars in luminosity—"yellow dwarf" indeed!

The red-dwarf stars offer us something new. When I discussed the displacement of the sun by Jupiter at the beginning of the article, I pointed out that the displacement would be larger, and therefore possible to observe from other stars, if Jupiter were considerably larger.

An alternative would be to have the sun considerably less massive. It is not the absolute mass of either component, but the ratio of the masses that counts. Thus, the Jupiter-sun ratio is 1:1000, which leads to an indetectable displacement. The mass ratio of the two components of the Sirius system, however, is 1:2.5, and that is easily detectable.

If a star were, say, half the mass of the sun, and if it were circled by a body eight times the mass of Jupiter, the mass ratio would be about 1:60. The displacement would not be as readily noticeable as in the case of Sirius, but it would be detectable.

Exactly such a displacement was detected in 1943 at

Sproul Observatory in Swarthmore College, in connection with 61 Cygni. From unevennesses in the motion of one of the major components, a third component, 61 Cygni C, was deduced as existing; a body with a mass $\frac{1}{125}$ of our sun or only eight times that of Jupiter. In 1960 similar displacements were discovered for the star Lalande 21185 at Sproul Observatory. It, too, had a planet eight times the mass of Jupiter.

And in 1963, the same observatory announced a third planet outside the solar system. The star involved is Barnard's Star.

This star was discovered in 1916 by the American astronomer Edward Emerson Barnard, and it turned out to be an unusual star indeed. In the first place it is the second nearest star to ourselves, being only 6.1 light-years distant. (The three stars of the Alpha Centauri system, considered as a unit, are the nearest, at 4.3 light-years; Lalande 21185 at 7.9 light-years is third nearest. Next is Wolf 359 and then the two stars of the Sirius system—at 8.0 and 8.6 light-years respectively.)

Barnard's Star has the most rapid proper motion known, partly because it is so close. It moves 10.3 seconds of arc a year. This isn't much, really, for in the forty-seven years since its discovery, it has only moved a little over 8 minutes of arc (or about one quarter the apparent width of the moon) across the sky. For a "fixed" star, however, that's a tremendously rapid movement; so rapid, in fact, that the star is sometimes called "Barnard's Runaway Star" or even "Barnard's Arrow."

Barnard's Star is a red dwarf with about one fifth the mass of the sun and less than $\frac{1}{2500}$ the luminosity of the sun (though it is nine times as luminous as Proxima Centauri).

The planet displacing Barnard's Star is Barnard's Star Band. It is the smallest of the three invisible bodies yet discovered. It is about $\frac{1}{700}$ the mass of the sun and hence roughly 1.2 times the mass of Jupiter. Put another way, it is about five hundred times the mass of the earth. If it possesses the over-all density of Jupiter, it would make a planetary body about 100,000 miles in diameter.

All this has considerable significance. Astronomers have about decided from purely theoretical considerations that

191

most stars have planets. Now we find that in our immediate neighborhood at least three stars have at least one planet apiece. Considering that we can only detect super-Jovian planets, this is a remarkable record. Our sun has one planet of Jovian size and eight sub-Jovians. It is reasonable to suppose that any other star with a Jovian planet has a family of sub-Jovians also. And indeed, there ought to be a number of stars with sub-Jovian planets only.

In short, on the basis of these planetary discoveries, it would seem quite likely that nearly every star has planets.

A generation ago, when it was believed that solar systems arose through collisions or near-collisions of stars, it was felt that a planetary family was excessively rare. Now we might conclude that the reverse is true; it is the truly lone star, the one without companion stars or planets, that is the really rare phenomenon.

And yet the red dwarfs aren't quite as little as they seem to be from their luminosity. Even the smallest red dwarf, Proxima Centauri, is not less than one tenth the mass of the sun. In fact, stellar masses are quite uniform; much more uniform than stellar volumes, densities, or luminosities. Virtually all stars range in mass from not less than one tenth of the sun to not more than ten times the sun, a stretch of but two orders of magnitude.

There is good reason for this. As mass increases, the pressure and temperature at the center of the body also increases and the amount of radiation produced varies as the fourth power of the temperature. Increase the temperature ten times, in other words, and luminosity increases ten thousand times.

Stars that are more than ten times the mass of the sun are therefore unstable, for the pressures associated with their vastly intense radiations blow them apart in short order. On the other hand, stars with less than one tenth the mass of the sun do not have an internal temperature and pressure high enough to start a self-sustaining nuclear reaction.

The upper limit is fairly sharp. Too-massive stars, except in very rare cases, blow up and actually don't exist. Too-light stars merely don't shine and can't be seen, so that the lower limit is an arbitrary one. The light bodies may exist even if they can't be seen.

Below the smallest luminous stars are, indeed, the non-luminous planets. In our own solar system, we have bodies up to the size of Jupiter, which is perhaps $\frac{1}{100}$ the mass of the feebly glowing Proxima Centauri. A body such as 61 Cygni C would have a mass one twelfth that of Proxima Centauri. Undoubtedly there must be bodies closing that remaining gap in mass.

Jupiter, large as it is for a planet, develops insufficient heat at its center to lend significant warmth to its surface. Whatever warmth exists on Jupiter's surface derives from solar radiation. The same may be true for 61 Cygni C.

However, as we consider planets larger still, there must come a point where the internal heat, while not great enough to start nuclear reactions, is great enough to keep the surface warm, perhaps warm enough to allow water to remain eternally in the liquid form.

We might call this a super-planet but, after all, it is radiating energy in the infrared. Such a body would not glow visibly, but if our eyes were sensitive to infrared we might see them as very dim stars. They might, therefore, be more fairly called "sub-stars" than super-planets.

Harlow Shapley, emeritus director of Harvard College Observatory, thinks it possible that such sub-stars are very common in space, and that they might even be the abode of life. To be sure, a sub-star with an earth-like density would have a diameter of about 150,000 miles and a surface gravity about eighteen times earth-normal. To life developing in the oceans, however, gravity is of no importance.

Is it possible that such a sub-star (with, perhaps, a load of life) might come rolling close enough to the solar system, some day, to attract exploring parties?

We can't be certain it won't happen. In the case of luminous stars, we can detect invaders from afar, and we can be certain that none will be coming this way for millions of years. A sub-star, however, could sneak up on us unobserved; we'd never know it was approaching. It might be right on top of us—say, within fifteen billion miles of the sun—before we detected its presence through its reflected light and through its gravitational perturbations on the outer planets.

Then at last mankind might go out to see for themselves

what a little star was like and set to rest that generations-long plaintive chant of childhood, "How I wonder what you are!"

Only—it won't be twinkling.

Part Six
GENERAL

17. The Isaac Winners

WHEN ONE LOOKS BACK over the months or years, it becomes awfully tempting to try to pick out the best in this or that category. Even the ancient Greeks did it, choosing the "seven wise men" and the "seven wonders of the world."

We ourselves are constantly choosing the ten best-dressed women of the year or the ten most notable newsbreaks, or we list the American Presidents in order of excellence. The FBI and other law-enforcement agencies even list criminals in the order of their desirability (behind bars, that is).

There is a certain sense of power in making such lists. An otherwise undistinguished person suddenly finds himself able to make decisions with regard to outstanding people, taking this one into the fold and hurling that one into the outer darkness. One can, after some thought, move x up the list and y down, possibly changing the people so moved in the esteem of the world. It is almost god-like, power like that.

Well, can I be faced with the possibility of assuming god-like power and not assume it at once? Of course not.

As it happens, I have been spending nearly two years writing a history of science, and in the course of writing it I could not help but grow more or less intimate with about a thousand scientists of all shapes and varieties.

Why not, then, make a list of the "ten greatest scientists of history"? Why not, indeed?

I sat down, convinced that in ten seconds I could rattle off the ten best. However, as I placed the cerebral wheels in gear, I found myself quailing. The only scientist who, it seemed to me, indubitably belonged to the list and who would, without the shadow of a doubt, be on such a list prepared by anyone but a consummate idiot, was Isaac Newton.

But how to choose the other nine?

It occurred to me to do as one did with the Academy Awards (and such-like affairs) and set up nominations, and after some time at that I found I had no less than seventy-two scientists whom I could call "great" with an absolutely clear conscience. From this list I could then slowly and by process of gradual elimination pick out my ten best.

This raised a side issue. I would be false to current American culture if I did not give the ten winners a named award. The motion picture has its Oscar, television its Emmy, mystery fiction its Edgar, and science fiction its Hugo. All are first names and the latter two honor great men in the respective fields: Edgar Allan Poe and Hugo Gernsback.

For the all-time science greats, then, why not an award named for the greatest scientist of them all—Newton. To go along with the Oscar, Emmy, Edgar, and Hugo, let us have the Isaac. I will hand out Isaac Awards and choose the Isaac winners.[1]

Here, then, is my list of nominees, with a few words intended to indicate, for each, the reasons for the nomination. These are presented in alphabetical order—and I warn you the choice of nominees is entirely my own and is based on no other authority.

1 *Archimedes* (287?–212 B.C.) Greek mathematician. Considered the greatest mathematician and engineer of ancient times. Discovered the principle of the lever and the principle of buoyancy. Worked out a good value for π

[1] If anyone has some wild theory that the choice of the name derives from any source other than Newton, let him try to prove it. Besides, what conceivable alternate origin could there be?

by the principle of exhaustion, nearly inventing calculus in the process (see Chapter 4).

2 *Aristotle* (384–322 B.C.) Greek philosopher. Codified all of ancient knowledge. Classified living species and groped vaguely toward evolutionary ideas. His logic proved the earth was round (see Chapter 15) and established a world system that was wrong, but that might have proved most fruitful if succeeding generations had not too slavishly admired him.

3 *Arrhenius, Svante A.* (1859–1927) Swedish physicist and chemist. Established theory of electrolytic dissociation, which is the basis of modern electrochemistry. Nobel Prize, 1903.

4 *Berzelius, Jöns J.* (1779–1848) Swedish chemist. Was the first to establish accurate table of atomic weights. Worked out chemical symbols still used in writing formulas. Pioneered electrochemistry and notably improved methods of inorganic analysis.

5 *Bohr, Niels* (1885–1962) Danish physicist. First to apply quantum theory to atomic structure, and demonstrated the connection between electronic energy levels and spectral lines. Suggested the distribution of electrons among "shells" and rationalized the periodic table of elements. Nobel Prize, 1922.

6 *Boyle, Robert* (1627–1691) Irish-born British physicist and chemist. First to study the properties of gases quantitatively. First to advance operational definition of an element.

7 *Broglie, Louis V. de* (1892–   ) French physicist. Discovered the wave nature of electrons, and of particles in general, completing the wave-particle duality. Nobel Prize, 1929.

8 *Cannizzaro, Stanislao* (1826–1910) Italian chemist. Established usefulness of atomic weights in chemical calculations, and in working out the formulas of organic compounds.

9 *Cavendish, Henry* (1731–1810) English physicist and chemist. Discovered hydrogen and determined the mass of the earth. Virtually discovered argon and pioneered in the study of electricity (see Chapter 11).

10 *Copernicus, Nicolaus* (1473–1543) Polish astronomer. Enunciated heliocentric theory of the solar system, with sun at center and earth moving about it as one of the

planets. Initiated the Scientific Revolution in the physical sciences (see Chapter 15).

11 *Crick, Francis H. C.* (1916–   ) English physicist and biochemist. Worked out the helical structure of the DNA molecule, which was the key breakthrough in modern molecular biology. Nobel Prize, 1962.

12 *Curie, Marie S.* (1867–1934) Polish-French chemist. Her investigations of radioactivity glamorized the subject. Discovered radium. Nobel Prize, 1903 (Physics) *and* 1911 (Chemistry). First person in history to win two.

13 *Cuvier, Georges L. C. F. D.* (1769–1832) French biologist. Founder of comparative anatomy and, through systematic studies of fossils, founder of paleontology as well.

14 *Dalton, John* (1766–1844) English chemist. Discovered law of multiple proportions in chemistry, which led him to advance an atomic theory that served as the key unifying concept in modern chemistry.

15 *Darwin, Charles R.* (1809–1882) English naturalist. Worked out a theory of evolution by natural selection which is the central, unifying theme of modern biology (see Chapter 13).

16 *Davy, Humphry* (1778–1829) English chemist. Established importance of electrochemistry by utilizing an electric current to prepare elements not previously prepared by ordinary chemical means. These included such elements as sodium, potassium, calcium, and barium.

17 *Ehrlich, Paul* (1854–1915) German bacteriologist. Pioneered in the staining of bacteria. Worked out methods of disease therapy through immune serums and also discovered chemical compounds specific against particular diseases, notably syphilis. Hence founder of both serum therapy and chemotherapy. Nobel Prize, 1908.

18 *Einstein, Albert* (1879–1955) German-Swiss-American physicist. Established quantum theory, earlier put forth by Planck, by using it to explain the photoelectric effect. Worked out the theory of relativity to serve as a broader and more useful world-picture than that of Newton. Nobel Prize, 1921.

19 *Faraday, Michael* (1791–1867) English chemist and physicist. Advanced the concept of "lines of force." Devised the first electric generator capable of converting mechanical energy into electrical energy. Worked out the

laws of electrochemistry and pioneered in the field of low-temperature work.

20 *Fermi, Enrico* (1901–1954) Italian-American physicist. Investigated neutron bombardment of uranium, initiating work that led to the atomic bomb, in the development of which he was a key figure. Outstanding theoretician in the field of subatomic physics. Nobel Prize, 1938.

21 *Franklin, Benjamin* (1706–1790) American universal talent. Demonstrated the electrical nature of lightning and invented the lightning rod. Enunciated the view of electricity as a single fluid, with positive charge representing an excess and negative charge a deficiency.

22 *Freud, Sigmund* (1856–1939) Austrian neurologist. Founder of psychoanalysis and revolutionized concepts of mental disease.

23 *Galileo* (1564–1642) Italian astronomer and physicist. Studied the motion of falling bodies, disrupting the Aristotelian world system and laying the foundation for the Newtonian one. He popularized experimentation and quantitative measurement and is the most important single founder of experimental science. He was the first to turn a telescope upon the heavens and founded modern astronomy.

24 *Gauss, Karl F.* (1777–1855) German mathematician and astronomer. Perhaps greatest mathematician of all time. In science, developed method of working out planetary orbit from three observations and made important studies of electricity and magnetism (see Chapter 5).

25 *Gay-Lussac, Joseph L.* (1778–1850) French chemist and physicist. Discovered several fundamental laws of gases and was the first to ascend in balloon to make scientific measurements at great heights.

26 *Gibbs, Josiah W.* (1839–1903) American physicist and chemist. Applied principles of thermodynamics to chemistry and founded, in detail, chemical thermodynamics, which is the core of modern physical chemistry.

27 *Halley, Edmund* (1656–1742) English astronomer. First to undertake systematic study of southern stars. Worked out the orbits of comets and showed that they were subject to the law of gravitation.

28 *Harvey, William* (1578–1657) English physiologist. First to apply mathematical and experimental methods to biology. Demonstrated the circulation of the blood, overthrowing ancient theories and founding modern physiology.

29 *Heisenberg, Werner* (1901–   ) German physicist. Enunciated uncertainty principle, a concept of great power in modern physics. Was the first to work out the proton-neutron structure of the atomic nucleus and was thus the founder of modern nucleonics. Nobel Prize, 1932.

30 *Helmholtz, Hermann L. F. von* (1821–1894) German physicist and physiologist. Advanced a theory of color vision and one of hearing, making important studies of light and sound. First to enunciate, clearly and specifically, the law of conservation of energy.

31 *Henry, Joseph* (1797–1878) American physicist. Devised first large-scale electromagnet and invented electric relay, which was basis of the telegraph. Invented the electric motor, which is the basis of much of modern electrical gadgetry.

32 *Herschel, William* (1738–1822) German-English astronomer. Discovered the planet Uranus, first to be discovered in historic times. Founded the modern study of stellar astronomy by work on double stars, on proper motions, etc. He was the first to attempt to work out the general shape and size of the Galaxy.

33 *Hertz, Heinrich R.* (1857–1894) German physicist. Discovered radio waves, thus establishing Maxwell's theoretical predictions concerning the electromagnetic spectrum (see Chapter 10).

34 *Hipparchus* (second century B.C.) Greek astronomer. The greatest of the naked-eye observers of the heavens. Worked out the epicycle theory of the solar system, with the earth at the center. Perfected system of latitude and longitude, devised first star map, and discovered the precession of the equinoxes (see Chapter 15).

35 *Hubble, Edwin P.* (1889–1953) American astronomer. His studies of the outer galaxies demonstrated that the universe was expanding. Presented first picture of known universe as a whole.

36 *Hutton, James* (1726–1797) Scottish geologist. Founded modern geology; the first to stress the slow, eons-long, changes of the earth's crust under environmental stresses continuing and measurable in the present.

37 *Huygens, Christian* (1629–1695) Dutch mathematician, physicist, and astronomer. Devised first pendulum clock, thus founding the art of accurate timekeeping. Im-

proved the telescope and discovered Saturn's rings. Was the first to advance a wave theory of light.

38 *Kekulé von Stradonitz, Friedrich A.* (1829–1896) German chemist. Devised the modern method of picturing organic molecules with bonds representing valence links, of which the carbon atom possessed four. This brought order into the jungle of organic chemistry.

39 *Kelvin, William Thomson, Lord* (1824–1907) Scottish physicist. Proposed absolute scale of temperature, did important theoretical work on electricity, and was one of those who worked out the concept of entropy.

40 *Kepler, Johann* (1571–1630) German astronomer. Established elliptical nature of planetary orbits, and worked out generalizations governing their motions. He thus established the modern model of the solar system and eliminated the epicycles that had governed astronomical thinking for nearly two thousand years.

41 *Kirchhoff, Gustav R.* (1824–1887) German physicist. Applied the spectroscope to chemical analysis, thus founding modern spectroscopy and laying the groundwork for modern astrophysics. He was the first to study blackbody radiation, something which led, eventually, to the quantum theory.

42 *Koch, Robert* (1843–1910) German bacteriologist. Isolated bacteria of tuberculosis and of anthrax. Was the first to develop systematic methods for culturing pure strains of bacteria and established rules for locating the infectious agent of a disease. Nobel Prize, 1905.

43 *Laplace, Pierre S.* (1749–1827) French mathematician and astronomer. Worked out the gravitational mechanics of the solar system in detail and showed it to be stable.

44 *Lavoisier, Antoine L.* (1743–1794) French chemist. First to popularize quantitative methods in chemistry. Established the nature of combustion and the composition of the atmosphere. Enunciated the law of conservation of matter. Introduced the modern system of terminology for naming chemical compounds and wrote the first modern chemical textbook (see Chapter 11).

45 *Lawrence, Ernest O.* (1901–1958) American physicist. Invented the cyclotron, first device suitable for induction of large-scale artificial nuclear reactions. Modern

nuclear-physics technology depends upon the cyclotron and its descendants. Nobel Prize, 1939.

46 *Leverrier, Urbain J. J.* (1811–1877) French astronomer. Worked out the calculations that predicted the position of the then-unknown Neptune. This was the greatest victory for gravitational theory and the most dramatic event in the history of astronomy.

47 *Liebig, Justus von* (1803–1873) German chemist. Worked out methods of quantitative analysis of organic compounds. Was the first to study chemical fertilizers intensively and hence is the founder of agricultural chemistry.

48 *Linnaeus, Carolus* (1707–1778) Swedish botanist. Painstakingly classified all species known to himself into genera, placed related genera into orders and related orders into classes, thus founding taxonomy. He devised the system of binomial nomenclature, whereby each species has a general and a specific name.

49 *Maxwell, James C.* (1831–1879) Scottish physicist. Worked out equations that served as basis for an understanding of electromagnetism. Showed light to be an electromagnetic radiation and predicted a range of such radiations beyond those then known. Worked out the kinetic theory of gases, one of the foundation blocks of physical chemistry (see Chapter 8).

50 *Mendel, Gregor J.* (1822–1884) Austrian botanist. His studies of pea plants founded the science of genetics, though the laws of inheritance he worked out remained unknown in his lifetime (see Chapter 13).

51 *Mendeléev, Dmitri I.* (1834–1907) Russian chemist. Worked out the periodic table of the elements, which proved an important unifying concept in chemistry. The value of the table was established by his dramatic prediction of the properties of as-yet-unknown elements.

52 *Michelson, Albert A.* (1852–1931) German-American physicist. Made accurate determinations of velocity of light. Invented the interferometer and used it to show that light travels at constant velocity in all directions despite motion of the earth. This served as the foundation of the theory of relativity (see Chapter 9). Nobel Prize, 1907.

53 *Moseley, Henry G. J.* (1887–1915) English physicist. Studied X-ray emission by elements and worked out the manner in which nuclear electric charge differed from element to element. This led to the concept of the atomic

number, which greatly improved the rationale behind the periodic table of the elements.

54 *Newton, Isaac* (1642–1727) English physicist and mathematician. Invented calculus, thus founding modern mathematics. Discovered compound nature of white light, thus founding modern optics. Constructed the first reflecting telescope. Worked out the laws of motions and the theory of universal gravitation, replacing Aristotle's world system with one that was infinitely better.

55 *Ostwald, Friedrich W.* (1853–1932) German physical chemist. Founder of modern physical chemistry. Worked on electrolytic dissociation. Proposed the modern view of catalysis as a surface phenomenon. Nobel Prize, 1909.

56 *Pasteur, Louis* (1822–1895) French chemist. Did pioneer work in stereochemistry. Advanced the germ theory of disease, thus founding modern medicine. He worked out dramatic methods of inoculation against various diseases.

57 *Pauling, Linus C.* (1901–   ) American chemist. Applied quantum theory to molecular structure, proposing a new and more useful view of the valence bond, and establishing modern theoretical organic chemistry. First to propose the helical structure of large organic molecules, such as proteins, which led on to Crick's work. Nobel Prize, 1954 (Chemistry) *and* 1963 (Peace). Second person to win two Nobel Prizes.

58 *Perkin, William H.* (1838–1907) English chemist. Initiated the great days of synthetic organic chemistry by synthesizing aniline purple, first of the aniline dyes. Also synthesized coumarin, founding the synthetic perfume industry.

59 *Planck, Max K. E. L.* (1858–1947) German physicist. Worked out quantum theory to explain the nature of black-body radiation. This theory treats energy as discontinuous and as consisting of discrete particles or quanta. The new understanding it offered is so crucial that physics is commonly divided into "classical" (before Planck) and "modern" (since Planck). Nobel Prize, 1918.

60 *Priestley, Joseph* (1733–1804) English chemist. Discovered oxygen (see Chapter 11).

61 *Roentgen, Wilhelm K.* (1845–1923) German physicist. Discovered X-rays, an event usually considered as initiating the Second Scientific Revolution. Nobel Prize, 1901.

62 *Rutherford, Ernest* (1871–1937) New Zealand-born

British physicist. Enunciated the theory of the nuclear atom, in which the atom was viewed as containing a tiny central nucleus surrounded by clouds of electrons. This founded subatomic physics. Rutherford was the first to effect an artificial nuclear reaction, changing one element into another. Nobel Prize, 1908.

63 *Scheele, Karl W.* (1742–1786) German-Swedish chemist. Discovered or co-discovered some half-dozen elements, as well as a variety of organic and inorganic compounds.

64 *Schwann, Theodor* (1810–1882) German zoologist. Discovered first animal enzyme, pepsin. Contributed to the disproof of spontaneous generation. Strongest single contributor to the establishment of the cell theory, which is virtually the atomic theory of biology.

65 *Soddy, Frederick* (1877–1956) English chemist. Worked out the isotope theory of the elements and with it the details of the course of radioactive breakdown. Nobel Prize, 1921.

66 *Thales* (640?–546 B.C.) Greek philosopher. Founder of rationalism and the tradition that has led to modern science.

67 *Thomson, Joseph J.* (1856–1940) English physicist. First to establish, definitely, that cathode rays consisted of particles far smaller than atom; therefore the discoverer of the electrons and the founder of the study of subatomic particles. Nobel Prize, 1906.

68 *Van't Hoff, Jacobus H.* (1852–1911) Dutch physical chemist. Advanced theory of the tetrahedral carbon atom, by which molecular structure could be described in three dimensions. Contributed greatly to chemical thermodynamics. Nobel Prize, 1901.

69 *Vesalius, Andreas* (1514–1564) Belgian anatomist. Described his anatomical observations in a book with classically beautiful illustration. This demolished ancient errors in anatomy and established the science in its modern form. Published in 1543 (the year of Copernicus' book) it began the Scientific Revolution in the biological sciences.

70 *Virchow, Rudolf* (1821–1902) German pathologist. Studied disease from the cellular standpoint and ranks as the founder of modern pathology. He also labored on behalf of sanitation reform and was one of the founders of modern hygiene.

71 *Volta, Alessandro* (1745–1827) Italian physicist. Built the first chemical battery and founded the study of current electricity.

72 *Wöhler, Friedrich* (1800–1882) First to form an organic compound (urea) from an inorganic precursor, thus founding modern organic chemistry.[2]

Having completed the list of nominees, I am under the temptation to play with it, analyze it statistically in various fashions. I shall succumb to this in only one small way. Let me list the total number of scientists on the list according (as nearly as I can guess) to the language they thought in.

English	26
German	21
French	7
Italian	4
Greek	4
Swedish	3
Dutch & Flemish	3
Polish	2
Danish	1
Russian	1

I suppose this can be taken as evidence that modern science is primarily an Anglo-American-German phenomenon. I think, though, it is more likely to demonstrate that the individual who selected the names is himself English-speaking.

[2]I repeat that this list is perhaps overconservative. Arguments can be advanced for including such men as Hippocrates, Euclid, Leonardo da Vinci, Robert H. Goddard, Charles H. Townes, Emil Fischer, and so on.

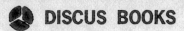 **DISCUS BOOKS**

DISTINGUISHED MODERN THEATRE AND FILM

ACTION FOR CHILDREN'S TELEVISION	10090	1.25
ANTONIN ARTAUD Bettina L. Knapp	12062	1.65
A BOOK ON THE OPEN THEATRE Robert Pasolli	12047	1.65
THE DISNEY VERSION Richard Schickel	08953	1.25
EDWARD ALBEE: A PLAYWRIGHT IN PROTEST Michael E. Rutenberg	11916	1.65
THE EMPTY SPACE Peter Brook	19802	1.65
THE EXPERIMENTAL THEATRE James Roose-Evans	11981	1.65
GREAT SCENES FROM THE WORLD THEATRE James L. Steffensen, Jr., Ed.	15735	1.95
GREAT SCENES FROM THE WORLD THEATRE, VOL. II James L. Steffensen, Jr., Ed.	12658	2.45
THE HOLLYWOOD SCREENWRITERS Richard Corliss	12450	1.95
INTERVIEWS WITH FILM DIRECTORS Andrew Sarris	21568	1.95
PICTURE Lillian Ross	08839	1.25
PUBLIC DOMAIN Richard Schechner	12104	1.65
THE RADICAL THEATRE NOTEBOOK Arthur Sainer	22442	2.25
SCENES FOR YOUNG ACTORS Lorraine Cohen, Ed.	14936	1.95
THE YOUNG FILMMAKERS Roger Larson, with Ellen Meade	08250	.95

Where better paperbacks are sold, or directly from the publisher. Include 25¢ per copy for mailing; allow three weeks for delivery.

Avon Books, Mail Order Dept.
250 West 55th Street, New York, N. Y. 10019

DMT 3-75

DISCUS BOOKS

DISTINGUISHED NON-FICTION

A SELECTION OF RECENT TITLES

DRT 3-75

DISCUS BOOKS

DISTINGUISHED NON-FICTION

American Civil Liberties Union Handbooks on The Rights of Americans

THE RIGHTS OF MENTAL PATIENTS
Bruce Ennis and Loren Siegel   10652   1.25

THE RIGHTS OF THE POOR
Sylvia Law   18754   .95

THE RIGHTS OF PRISONERS
David Rusovsky   07591   .95

THE RIGHTS OF SERVICEMEN
Robert S. Rivkin   07500   .95

THE RIGHTS OF STUDENTS
Alan H. Levine and Eve Cary   05776   .95

THE RIGHTS OF SUSPECTS
Oliver Rosengart   18606   .95

THE RIGHTS OF TEACHERS
David Rubin   07518   .95

THE RIGHTS OF WOMEN
Susan Deller Ross   17285   1.25

THE RIGHTS OF REPORTERS
Joel M. Gora   21485   1.25

THE RIGHTS OF HOSPITAL PATIENTS
George J. Annas   22459   1.50

Wherever better paperbacks are sold, or directly from the publisher. Include 25¢ per copy for mailing; allow three weeks for delivery.

Avon Books, Mail Order Dept.
250 West 55th Street, New York, N.Y. 10019